Advances in Marine Biology

This is a numbered series focused on the latest research in marine biology.

nova
science publishers
www.novapublishers.com

Advances in Marine Biology

Advances in Marine Biology. Volume 6
Adam Kovács and Patrik Nagy (Editors)
2021. ISBN: 979-8-88697-996-1 (Hardcover)
2021. ISBN: 979-8-89113-021-0 (eBook)

Advances in Marine Biology. Volume 5
Adam Kovács and Patrik Nagy (Editors)
2021. ISBN: 978-1-68507-066-3 (Hardcover)
2021. ISBN: 978-1-68507-080-9 (eBook)

Advances in Marine Biology. Volume 4
Adam Kovács and Patrik Nagy (Editors)
2019. ISBN: 978-1-53615-042-1 (Hardcover)
2019. ISBN: 978-1-53615-045-2 (eBook)

Advances in Marine Biology. Volume 3
Adam Kovács and Patrik Nagy (Editors)
2018. ISBN: 978-1-53613-092-8 (Hardcover)
2018. ISBN: 978-1-53613-093-5 (eBook)

Advances in Marine Biology. Volume 2
Adam Kovács and Patrik Nagy (Editors)
2017. ISBN: 978-1-53612-749-2 (Hardcover)
2017. ISBN: 978-1-53612-750-8 (eBook)

Advances in Marine Biology. Volume 1
Adam Kovács and Patrik Nagy (Editors)
2016. ISBN: 978-1-53610-131-7 (Hardcover)
2016. ISBN: 978-1-53610-148-5 (eBook)

More information about this series can be found at
https://novapublishers.com/product-category/series/advances-in-marine-biology/

Adam Kovács
and Patrik Nagy
Editors

Advances in Marine Biology

Volume 6

nova
science publishers

www.novapublishers.com

NOTICE TO THE READER

Library of Congress Cataloging-in-Publication Data

ISBN: 979-8-88697-996-1

Published by Nova Science Publishers, Inc. † New York

Contents

Preface

This volume contains six chapters that detail recent advances in marine biology. Chapter One highlights the usefulness of zebrafish for biological screening of hydrogels based on amino acid and acrylates and hydrogel scaffolds based on alginate, gelatin and 2-hydroxyethyl methacrylate. Chapter Two aims to critically review the effects of climate change on green mussels and the environmental factors that may influence mussel production. Chapter Three explores how certain oceanographical phenomenas or events affect local marine ecosystems and the cetacean communities in Indonesia. Chapter Four assesses the bycatch and discardment in benthic trawling in Northern Tunisia. Chapter Five covers the parasitic diseases that are common in rainbow trout, the ways to prevent and control these diseases, the treatment methods of the diseases, and the life cycles of the parasites. Lastly, Chapter Six reveals a comprehensive study of the nutritional content, fatty acid profile and shelf life of rainbow trout.

Chapter 1

Studies of Zebrafish and Their Behavior for (Meth)Acrylate-based Hydrogels and Hydrogel Scaffolds

Simonida Lj. Tomić[1,*]
Jasmina Runić-Nikodinović[2]
Ivana Aleksić[2]
Marija Vukomanović[3]
Vuk V. Filipović[4]
and Marija M. Babić Radić[1]
[1]University of Belgrade, Faculty of Technology and Metallurgy, Belgrade, Serbia
[2]University of Belgrade Institute of Molecular Genetics and Genetic Engineering, Belgrade, Serbia
[3]Jožef Stefan Institute, Advanced Materials Department, Ljubljana, Slovenia
[4]University of Belgrade, Institute for Chemistry, Technology and Metallurgy, Belgrade, Serbia

Abstract

Hydrogels and hydrogel scaffolds as polymeric biomaterials have gained great importance and potency in the medical and pharmaceutical fields. However, these types of polymeric biomaterials can cause negative effects on living species during usage. Thus, it is essential to identify hydrogels and hydrogel scaffolds with the least adverse effects on aquatic

---

[*] Corresponding Author's Email: simonida@tmf.bg.ac.rs.

In: Advances in Marine Biology. Volume 6
Editors: Adam Kovács and Patrik Nagy
ISBN: 979-8-88697-996-1

organisms. To determine the toxicity of these polymeric biomaterials, we investigated the effects of hydrogels based on amino acid and acrylates and hydrogel scaffolds based on alginate, gelatin and 2-hydroxyethyl methacrylate on zebrafish embryos using the fish embryo toxicity test (FET). In summary, all tested hydrogels and hydrogel scaffolds displayed the least toxic impact, thus suggesting adequate environmental biocompatibility for the generation of hydrogels and hydrogel scaffolds. This review highlights the usefulness of zebrafish for biological screening of hydrogels based on amino acid and acrylates and hydrogel scaffolds based on alginate, gelatin and 2-hydroxyethyl methacrylate.

Keywords: polymers, hydrogels, scaffolds, alginate, gelatin, 2-hydroxyethyl methacrylate, apatite, *in vivo* biocompatibility, zebrafish

Introduction

Recent scientific studies have been indicated that animal models with the ability to regenerate tissue are more suitable options for regenerative medicine studies in several types of diseases, compared to models without regenerative abilities (Zang et al. 2018; Zang et al. 2015; Usai et al. 2020). Hence, the zebrafish has been introduced as a model that possesses the capacity to regenerate different organs and tissues, and fine regeneration in zebrafish has been broadly investigated in the regenerative medicine field. Additionally, the zebrafish is a suitable model for studying a variety of different situations. It has been used for developmental studies because of the special characteristics of its larva. The features of the zebrafish make it a desirable animal model, the advantages of zebrafish and recent research shows that zebrafish is a promising animal model for personalized developmental and degenerative diseases. It turns out that zebrafish can have a leading role in regeneration studies of endocrine diseases and provide a good perception of underlying mechanisms (Arjmand et al. 2020).

Animal testing has long been used in scientific researches to study complex biological phenomena that cannot be investigated using two-dimensional cell cultures in plastic dishes. With time, it appeared that more differences could exist between cell cultures and animal models and even more when translated to human patients. Innovative models became essential for developing more accurate knowledge. Tissue engineering can be used to provide some of those models, but it mostly relies on the use of prefabricated scaffolds on which cells are seeded. The self-assembly protocol has recently produced organ-specific human-derived three-dimensional models without the need for exogenous material. This strategy will help to achieve the 3R principles.

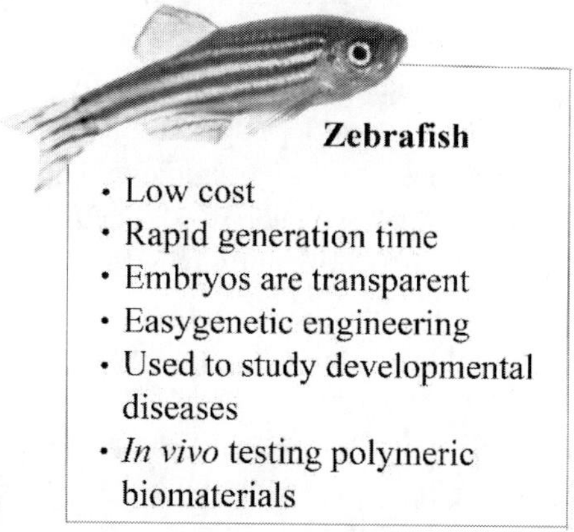

Scheme 1. Advantages and applications of the zebrafish model in biomedical research.

Zebrafish are selected as an important vertebrate model organism for developmental biology, neurobiology and toxicology due to their small size, rapid generation time and optically transparent embryos (Xu and Zon 2010; Kabashi et al. 2011). Zebrafish embryos can develop to fully functioning juveniles by 120 h, which allows for a high-throughput screen of developmental related disease at low cost (George et al. 2011). These zebrafish-specific advantages confer an ideal model for rapid and large-scale toxicity screens (Scheme 1).

Zebrafish, a vertebrate model, presents specific properties that make it a suitable choice for investigating in different fields of biomedicine. Unlike several animal models, the zebrafish has the ability to regenerate its organs, such as fins, central nervous system (CNS), heart, pancreas, liver, and kidney. Regarding this given capacity, it can be used for different models of injury in cardiovascular, neurological, and metabolic diseases (Zang et al. 2018; Zang et al. 2015). Additionally, the zebrafish is used as an animal model in personalized medicine for studying different diseases like cancers, as avatars due to its beneficial advantages (Usai et al. 2020; Costa et al. 2020). Moreover, zebrafish were used to investigate one of the unique features in humans, the gut microbiota, to find the associations between metabolic disorders and intestinal microbiota (Faillaci et al. 2018).

Evaluation of toxicity of chemicals, compounds and/or materials is the most important step in researches, which can lead to the discovery of a new agent(s), that can be used in clinics for the treatment of a variety of diseases. *In vitro* studies with human cells and tissues can be very useful for metabolism and mode of action studies (MacGregor et al. 2001), while *in vivo* experiments on live animals can provide considerably more information including pharmacokinetics, impact on development and specific effects on different organs. Although limitations of *in vivo* studies where animal trials did not predict severe human toxicity have been recognized (Van Norman 2019), the use of animals in science has significantly benefitted humans. In this review, we will highlight the importance and advantages of the small freshwater fish, *Danio rerio*, and important studies on the toxicity of new biomaterials for this model system.

Zebrafish was first described by a Scottish physician Francis Hamilton in 1822, found in the river Ganges near the state of Bihar in north-eastern India (Hamilton 1822). The name of the species has undergone a few changes to its current Latin name *Danio rerio* (some recent synonyms were *Cyprinus* and *Brachydanio*). The first time zebrafish was mentioned in the context of research was in 1972 by American molecular biologist George Streisinger, who suggested zebrafish as a model for studying vertebrate development due to several important qualities. These qualities include the production of a large number of embryos (50 to 1000 per fertilisation) accessible at the earliest stages of development, the optical clarity of the developing fish, and a diploid genome allowing genetic studies (Houart 2001). In comparison to other *in vivo* model systems currently in use, the zebrafish has numerous advantages including a fully sequenced genome, high fecundity, short generation time, rapid embryonic development (24 h), and external fertilization (Teame et al. 2019). Unlike other animal models, zebrafish embryos and the fish in the early stages of development are fully translucent, and all organs can easily be visualized using a simple optical microscope. Blood circulation and vessels are also fully visible and approachable for laboratory manipulation. Furthermore, the maintenance of these animals in the laboratory is affordable, doesn't take a lot of laboratory space and growing of a reproductively capable male/female is less than 3 months under appropriate conditions (Teame et al. 2019). There are many different strains of zebrafish available for research, and some of the commonly used in biomedical research are AB, Casper, Ekkwill, Nadia, Wild Indian Karyotype, wild-caught, and Tubingen. Zebrafish strains are available for research purposes through the Zebrafish information link, a database of genetic and genomic data (https://zfin.org/action/feature/wildtype-list).

Zebrafish is an extensively studied vertebrate model organism, judged by the steadily increasing number of scientific publications over the last 20 years (Figure 1). New techniques have unquestionably contributed to the rising popularity of this model, with some of the latest research on zebrafish, based transcriptomics (Silva et al. 2021), mRNA tracking (Holler et al. 2021), CRISPR-Cas9 (Zhang et al. 2018) and advanced microscopy (Høgset et al. 2020). Results from zebrafish studies have implications on numerous human diseases, including neural disorders, cancer infectious diseases, cardiovascular diseases, kidney diseases, diabetes (Zang, Shimada and Nishimura 2018), blindness (Leyk et al. 2017), deafness (Cairns et al. 2021), hematopoiesis (Chalin et al. 2021), and muscle disorders (Kürekçi et al. 2021). Human diseases studies on zebrafish are considered approximately accurate giving the fact that by direct comparison of the zebrafish and human protein-coding genes reveals that 71.4% of human genes have at least one zebrafish orthologue (Howe et al. 2013).

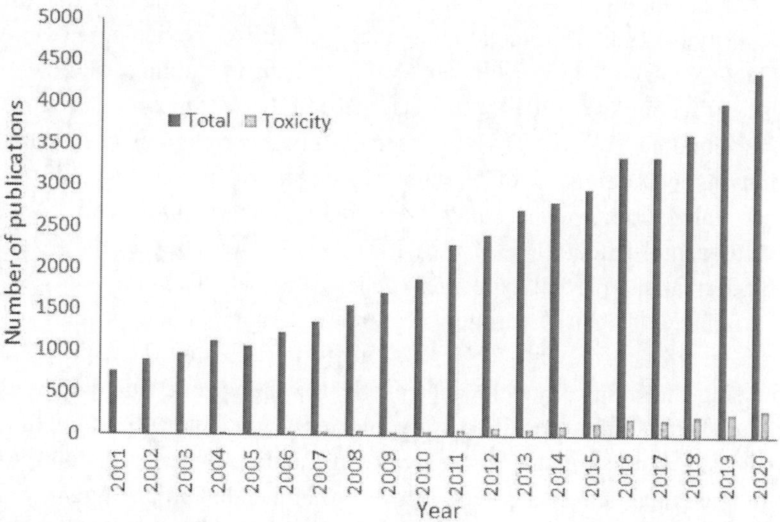

Source PubMed database.

Figure 1. The number of scientific publications with zebrafish as a model system in general and zebrafish toxicity.

Although only a fraction (up to 7%) of the total number of published papers regarding zebrafish, publications describing toxicity studies using this model are also increasing (Figure 1). The zebrafish developmental toxicity assay has recently become a mainstay of high-throughput analyses. It should also be noted that zebrafish are also being utilized in high-throughput screening for drug discovery (Wiley, Redfield and Zon 2017), both for the discovery of new compounds and the identification of new, unrecognized targets (MacRae and Peterson 2015; Patton, Zonand Langenau 2021).

In developmental toxicity zebrafish assays (Buschmann 2013), generally multiple zebrafish embryos (20 to 50) are exposed to a range of chemicals and/or range of concentrations for five days, until the full organ development. Embryos at the 6 hours post-fertilization (hpf) stage are suitable for treatment (Kimmel et al. 1995). Toxicity is usually monitored by evaluating lethality, teratogenicity, or other phenotypic changes (Table 1), according to Organisation for Economic Co-operation and Development (OECD) guidelines for the testing of chemicals Test 236 (OECD 2013). At the end of the exposure period, acute toxicity is determined based on a positive outcome in any of the observations recorded, and the LC50 is calculated (Buschmann 2013).

These toxicity assays became also a standard component of the United States Environmental Protection Agency's (US EPA) Toxicology Testing in the 21st Century (Tox21), both ToxCastTM Phase I (Padilla et al. 2011) and Phase II (Truong et al. 2014; Volz et al. 2015). It was shown that the results of the zebrafish ToxCast TM screens show good correlations to mammalian toxicity assays. Zebrafish LC50 values for acute toxicity correlating with rat LC50 inhalation values, and zebrafish Lowest Observed Adverse Developmental Effect Dose (LOADED) values correlating to the results of rabbit dermal and rat oral exposures (Ducharme et al. 2015).

Alginate/collagen hydrogels containing therapeutic, pro-angiogenic *fibroblast growth factor 2* (FGF-2), and formulated as microspheres, is a promising and clinically relevant vehicle for therapeutic angiogenesis. By titrating the amount of readily dissolvable and degradable collagen with more slowly degradable alginate in the hydrogel mixture, the degradation rates of the biomaterial controlled release kinetics of embedded proangiogenic FGF-2 can be adjusted. Furthermore, elaboration of a microsphere synthesis protocol allowing accurate control over sphere size, also a critical determinant of degradation/release rate. As expected, alginate/collagen microspheres were completely biocompatible and did not cause any adverse reactions when injected in mice. Importantly, the amount of pro-angiogenic FGF-2 released from such microspheres led to robust induction of angiogenesis in zebrafish embryos similar to that achieved by injecting FGF-2-releasing cells. These findings highlight the use of microspheres constructed from alginate/collagen hydrogels as a promising and clinically relevant delivery system for pro-angiogenic therapy (Ali et al. 2018; Lawson and Weinstein 2002; Lee et al. 2009; Rouhi et al. 2010).

Table 1. Lethal and teratogenic effects observed in zebrafish embryos at different hours post-fertilization (hpf)

Category	Developmental endpoints	Exposure time (hpf)			
		24 h	48 h	72 h	96 h
Lethal effect	Egg coagulation[a]	•	•	•	•
	Lack of heart-beat		•	•	•
Teratogenic effect	Malformation of head	•	•	•	•
	Malformation of eyes[b]	•	•	•	•
	Malformation of sacculi/otoliths[c]	•	•	•	•
	Malformation of chorda	•	•	•	•
	Malformation of tail[d]	•	•	•	•
	Scoliosis	•	•	•	•
	Yolk edema	•	•	•	•
	Yolk deformation	•	•	•	•
	Growth retardation		•	•	•
	Hatching			•	•
	Growth retardation[e]	•	•	•	•

[a] No clear organs structure is recognized

[b] Malformation of eyes was recorded for the retardation in eye development and abnormality in shape and size.

[c] Presence of no, one or more than two otoliths per sacculus, as well as reduction and enlargement of otoliths and/or sacculi (otic vesicles).

[d] Tail malformation was recorded when the tail was bent, twisted or shorter than to control embryos as assessed by optical comparation.

[e] Growth retardation was recorded by comparison with the control group of embryos in regards to development or size (before hatching, at 24 hpf and 48 hpf) or body length (after hatching, at and onwards 72 hpf).

Using the brain injury model of adult zebrafish, it was demonstrated that the damaged mobile function of injured zebrafish was better rescued by neural stem cells (NSCs) embedded in the chitosan-cellulose nanofiber (CS–CNF2) hydrogel that had better self-healing properties. We speculated that the CS–CNF2 hydrogel injected in the injured zebrafish may heal more quickly to occupy a proper space for tissue remodeling. Hydrogel injection was reported to facilitate the remodeling of myocardial infarcted areas and subsequent tissue repair (Lu et al. 2009; Johnson and Christman 2013). Brain injury also requires space for tissue remodeling (Cheng et al. 2005). The excellent space remodeling effect as well as the facilitated nutrient transport may contribute to the positive correlation between the self-healing property and neural regeneration by the hydrogels.

The influence of hydrophobically substituted quaternized hydroxyethyl cellulose polymers (SL) on the acute ecotoxicity of organisms living in the aquatic environment was studied (Simões et al. 2021). The ecotoxicity of four SL samples with different degrees of hydrophobic substitution (SL-5, SL-30, SL-60, SL-100 in mg L^{-1}) was investigated for seven species: *Vibrio fischeri*, *Raphidocelis subcapitata*, *Chlorella vulgaris*, *Daphnia magna*, *Brachionus calyciflorus*, *Heterocypris incongruens,* and *Danio rerio*. Hazard concentrations were derived using mean effective concentrations, realized from species-sensitive distribution curves. The influence of particle size, zeta potential and rheological properties were investigated. According to the results, the relatively low zeta potential of SL made them unstable in suspension. *Raphidocelis subcapitata*, *C. vulgaris* and *B. calyciflorus* were the most sensitive to all hydrophobically substituted samples indicating a possible imbalance of the lowest trophic levels. Accordingly, SL-5 is suggested as the eco-friendliest and is recommended to be used in the production of care products, to the detriment of the other three tested variants. The lowest level of hydrophobic substitution was found to be the least toxic variation.

A new method for harmful algae removal based on CaO_2@PEG- loaded water-soluble self-branched chitosan was investigated (Lin et al. 2021). This new method is much simpler than the complicated treatments and costly instruments in commercial use. The one-step removal of algae from water by (CP-SBC) system can be performed without additional instrumentation demonstrating better flocculation performance than commercial flocculants. This is due to self-branched chitosan flocculation properties (induced by CaO_2@PEG) because of the enhanced bridging and sweeping effect of branched chitosan. Zebrafish test and algae activity test showed remarkable biocompatibility of CP-SBC. CaO_2@PEG-loaded self-branched chitosan can be used to spontaneous float the flocs after flocculation by sustainably released O_2, as an "Air flotation system". Besides, CP-SBC can improve water quality by minimizing dissolved oxygen consumption and reducing total phosphorus concentrations.

Polyethyleneimine (PEI) is a positively charged polymer very important as an effective *attachment factor* for weakly anchoring cell lines and primary cells. PEI coatings were studied as materials capable to increase the ability of some cell lines and primary cultures to attach to culture plates (Vancha et al. 2004). PC-12 and HEK-293 cells plated on dishes coated with poly-ethyleneimine showed strong adhesion and homogeneous distribution. It is important to emphasize that mechanical interaction between a cell and its extracellular matrix polymers, such as polyethyleneimine, influence and control cell behavior and function. PEI can be used to increase adherence and allow axonal outgrowth from zebrafish retinal explants. Studying the effects of PEI coating on the transfection of loosely attaching cell lines it was determined that polyethyleneimine affords strong attachment of weakly anchoring cell lines and primary cells. When PEI is applied for lipofection protocols the procedure is more reliable and the yield of expressed products with PC-12 and HEK-293 cells is increased.

A method to investigate the biocompatibility of 3D printed devices by monitoring the viability and development of zebrafish embryos overextended culture was developed (Macdonald et al. 2016). The authors concluded that many of the untreated photopolymers used in commercial 3D printers are harmful to zebrafish culture. So VisiJet Crystal, classified as a USP Class VI material that was previously certified for use in animals, did not show favourable biocompatibility in zebrafish embryos test.

Pretreating photopolymer Fototec 7150 upgraded its compatibility with zebrafish culture, and it has been confirmed that pretreating photopolymer Fototec 7150 microfluidic devices can be used for biological applications (Macdonald et al. 2016). According to the results of this investigation, it is essential to test the biocompatibility of each photopolymer to enable biocompatible products in the 3D printing of microfluidic devices. The authors describe a method of treatment of the microfluidic devices which allows preserving zebrafish embryos development.

A one-step solid dispersion method was applied to prepare quercetin (Que) micelles using calculated amounts of Que and poly(ethylene glycol) monomethyl ether-ε-caprolactone copolymer (MPEG–PCL) (Wu et al. 2013). Micelles were investigated for their therapeutic effect on both cisplatin-sensitive (A2780s) and cisplatin-resistant (A2780cp) human ovarian cancer models. In the transgenic zebrafish models, stronger inhibitory effects of Que micelles were observed on embryonic angiogenesis, tumor-induced angiogenesis, tumor growth, and tumor metastasis. Furthermore, in comparison with free Que, Que micelles had improved anti-tumor and anti-metastasis activities.

The chetomin loaded micelles (Che-M) were prepared from monomethyl poly(ethylene glycol)-ε-caprolactone copolymer (PEG–PCL–PEG) and chetomin (1:19 wt ratio) using a solid dispersion method (Wu et al. 2014). Micelles were obtained as an aqueous formulation to outbalance the hydrophobicity of chetomin. Che-M micelles can also form a thermosensitive hydrogel (Che-H) at copolymer concentrations higher than the critical gelation concentration. Che-M showed equal cytotoxicity with free chetomin, but the apoptosis-inducing effects of Che-M were more significant. In transgenic zebrafish models, Che-M dramatically inhibited embryonic angiogenesis, tumor-induced angiogenesis and tumor growth.

The zebrafish *in vivo* model was used to evaluate the effects of PEG-b-PCL nano-micelle on cardiovascular system development (Zhou et al. 2016). Although biodegradable poly(ethylene glycol) (PEG) and poly(ε-caprolactone) (PCL) (PEG-b-PCL) based nanomicelles, have been extensively studied since they are widely used as drug carriers, their impact on the cardiovascular system has not been sufficiently examined. Therefore, it is of great importance to investigate if PEG-b-PCL nano-micelles have a harmful influence on the cardiovascular system. The results of the zebrafish model investigation showed that PEG-b-PCL nano-micelle can trigger off embryo mortality just as embryonic and larval malformations depending on the dose administered. It can be concluded that this study indicates that polymeric PEG-b-PCL nano-micelle could appoint eventual hazards to cardiovascular development.

Macrophage cells are an integral factor in maintaining a stable state of the innate immune system. The potential to transform disease treatment using nanoparticles (NPs) tailored to accurately target macrophages have a great potential in chronic inflammatory diseases, cancer treatment as well as vaccine development. In the research of Crecente-Campo et al. (2019), authors try to elucidate the interaction of nanocapsules (NCs) with macrophages as a function of NCs size and shell composition, both *in vitro* and *in vivo*. Two different sizes of inulin and chitosan NCs were used to clarify the role of the size and shell composition of NCs in interaction with macrophages. Small NCs interacted more efficiently with macrophages than the larger ones, from which inulin NCs were significantly less toxic than chitosan NCs. Following *in vivo* administration (intravenous/intramuscular) to zebrafish, small NCs, regardless of their composition, spread out considerably faster and further than their medium size NCs.

To enable a scientific approach for the design and optimization of nanomedicines Tao and collaborators (Tao et al. 2020) synthesized (OPMs) micelles using triblock poly(ethylene oxide)/poly(propylene oxide) copolymer (PEO-PPO-PEO), and disulfide bond crosslinked PEO-PPO-PEO copolymer (CPMs) to *measure* their elimination mechanisms *in vitro* and *in vivo*. Their integrity *in vivo* was visually investigated in zebrafish larvae across the entire living organism and at the cellular molecular level after intravenous microinjection. OPMs were rapidly disassociated and eliminated from blood circulation. CPMs were more stable and had a longer circulation in blood. Using the zebrafish model the authors were first to discover that disassociated polymeric micelles were eliminated through the hepatobiliary system. On the other hand, the CPM micelles were relatively difficult to eliminate.

Since its first report in the 1960s by Wichterle and Lim, hydrogels are increasingly being used as attractive and promising candidates for various biomedical applications. Thus, the assessment of the biological safety and possible biomedical application of hydrogels has become of primary importance for the suitable development of these biomaterials. Various *in vitro* and *in vivo* methods, due to their ethical and economical aspects are encouraged for the testing of new materials but despite that *in vivo* models are still needful. In the last decade, the use of zebrafish models significantly increased in different biomedical researches as well as for the faster and more economic testing of new biomaterials (Rothenbucher et al. 2019). Zebrafish have served as a model to evaluate the biological safety and biocompatibility of different synthetic and natural hydrogels for biomedical applications. Azman et al. synthesized the *Dioscorea hispida* starch-grafted polyacrylamide hydrogels by chemical polymerization method in basic solution and evaluated their beneficial use as disinfectants and biomaterials (Azman et al. 2015). The zebrafish embryo model was used to evaluate the biocompatibility of the *D. hispida* starch-grafted polyacrylamide hydrogel. Zebrafish embryo toxicity tests were performed as outlined in the Organisation for Economic Co-operation and Development (OECD) draft guidelines (Azman et al. 2015). The obtained data demonstrated that the 1:2 polyacrylamide-starch stock solution gave the highest embryo survival rate in comparison with 2:1 and 3:5 solutions; 0% of the embryos survive after 96-h exposure to the undiluted solution, while 88.89%, 77.89%, and 88.89% of embryos exposed to 1:100, 1:1000, and 1:10,000 test solutions survived after 96 h of observation and the zebrafish embryos were developing normally and displayed no signs of deformation that could be caused by the test solutions thus the hydrogel with 2:1 polyacrylamide: starch ratio shows an acceptable level of toxicity and excellent biocompatibility for biomedical application (Azman et al. 2015).

The polymeric drug carrier, which improves the uptake of the model drug valproate through the gastrointestinal tract, was synthesized by crosslinking of maleimide modified dextran (dextran maleimide), followed by reacting dextran maleimide with poly(ethylene glycol) dithiol (Gao 2018). Finally, using β-cyclodextrins conjugated to a dextran-PEG network, cyclodextrins were conjugated to the dextran backbone to provide drug binding cavities. Toxicity assay using zebrafish embryos showed that by applying this macromolecular drug carrier the sensitivity of the zebrafish embryotoxicity assay toward the model drug valproate is significantly enhanced.

Hydrogels with self-healing behavior and the ability to withstand multiple damage-healing cycles were prepared by Balitaan et al. (2020), using acrylamide-modified β-chitin (Am- β-Chn) derivatives cross-linked with alginate dialdehyde derivates (ADA). These hydrogels were obtained from sustainable natural resources and had excellent wound healing properties. The zebrafish model was used to examine how these hydrogels affect the healing of full-thickness cutaneous wounds. Indirect application of the hydrogels on full-thickness wounds of zebrafish led to significantly accelerated wound contraction.

An advanced hydrogel surface modification with bioactive molecules by modifying poly(ethylene glycol) (PEG)-based hydrogel (poly trimethylene carbonate)/acryloyl cyclic carbonate-PEG-poly trimethylene carbonate/ acryloyl cyclic carbonate (PTMAc-PEG-PTMAc, PTAE) copolymer) as a carrier of neural stem cells for the treatment of central nervous system disorders was achieved (Yang et al. 2017). The copolymer was modified by Arg-Gly-Asp-Cys (RGDC) and hyaluronic acid (HA). The best results were obtained with the Gel-1 (30% RGDC and HA 33%), which increased the survival and differentiation rates of NSCs and showed low toxicity to NSCs, zebrafish embryogenesis, as well as the liver and kidneys in rats with spinal cord injuries at 1, 4, and 8 weeks.

The self-healing hydrogel was prepared for central nervous system treatment in the form of the semi-interpenetrating polymeric network of chitosan, using hyaluronan (HA) as an interpenetrant (Liu et al. 2020). These self-healing hydrogels are meant to be injected locally to fill tissue defects after gelation. *The healing efficiency* of hydrogels was investigated by the zebrafish traumatic brain injury (TBI) and rat intracerebral hemorrhage (ICH) models. Both models confirmed good biocompatibility and therapeutic effect of CH hydrogel. The TBI zebrafish model also revealed a better repair function of chitosan-HA hydrogel in comparison with the chitosan hydrogel. Besides, the chitosan-HA hydrogel induced lesser extents of the inflammatory responses (edema, leukocyte infiltration, microgliosis, and astrogliosis) and glial scar formation after implantation into the rat brain compared to the chitosan hydrogel.

Based on recent literature and the importance of translating scientific results regarding novel biomedical materials into the commercial sector, the increase of works describing the toxicity of materials using the zebrafish model is expected to rise. The most commonly analysed materials include nanomaterials, biopolymers and hydrogels (Table 2).

The expansion of nanotechnology resulted in the generation of a variety of nanoparticles suitable for biomedical applications (McNamara and Tofail 2017). More recent studies implicate that inadequate usage of gold and silver nanoparticles can lead to potential risks for human health and ecosystems. Recently a review has been published with collated data regarding plasmonic nanoparticles toxicity in the zebrafish model (d'Amora et al. 2021). Overall, it was shown that gold nanoparticles induced minimal or low toxicity, characterized by small changes in the different biological parameters examined in embryos and larvae. Adults presented signs of genotoxicity and damage at the histological level in the ovary and liver. On the other hand, silver nanoparticles caused strong developmental toxicity, including mortality, malformations in organs and behavioral changes. These findings were supported by numerous previously published works, where it can be concluded the greater toxicity of silver nanoparticles vs. gold ones.

Apart from metal-based nanoparticles, numerous studies covered the toxicity of coated nanoparticles including lignin NPs (Stine et al. 2021). Stine and collaborators tested chitosan-coated lignin NPs and proved that lignin NPs did not affect zebrafish development while coated with chitosan induced severe toxicity in direct contact in higher doses >650 mg/l (Stine et al. 2021). Nanoparticles with silica coating (SiNPs) were also examined in the direct assay in serial concentrations, where it was found that in lower concentrations (up to 50 μg/mL) there was no lethal effect but with the increased dosages mortality has increased significantly (Duan et al. 2013). On the other side graphene-coated nanomaterials when tested on zebrafish, damaged zebrafish embryos by influencing the embryo hatching rate and body length in a concentration-dependent manner. Although there were no obvious effects on embryo development, graphene was adhered to a surface of chorion on some tested embryos, wrapping them and causing hypoxia and hatching delay. In further development graphene aggregated in organs (eyes, heart, yolk sac etc.) which caused apoptosis and reactive oxygen species (ROS) (Ou et al. 2016).

Zebrafish model was used for toxicity evaluation of biodegradable and biocompatible polymer polyhydroxyalkanoates (PHA) (Li et al. 2016). This extensive study among other findings confirms that PHA is a safe, non-toxic material with no significant embryonic abnormality or mortality rates detected. Noteworthy is the usage of biopolymers for microinjecting into embryos. The authors of this paper microinjected the PHA solutions into embryos at the 1-4 cell stage to evaluate the effect of PHAs on the early development of embryonic and larval zebrafish and the behavioral effect of PHAs. While PHA did not have any toxic effect on the development of zebrafish, there was noted some behavioural changes and colour preference behaviours of zebrafish larvae.

Table 2. Scientific research publications describing the toxicity of materials using the zebrafish model

Material	Toxicity	Reference
TiO2/Au nanoparticles	Not toxic	Brundo et al. 2016
Lignin nanoparticles (LNP) coated with chitosan	Toxic in higher doses >650 mg/l	Stine et al. 2021
Silver and gold nanoparticles	Au nanoparticles induce low toxicity while Ag nanoparticles caused strong developmental toxicity	d'Amora et al. 2021
Silica nanoparticles	A lower dose (25 and 50 mg/mL SiNPs) were not toxic while higher doses (100 and 200 mg/mL SiNPs) were significantly toxic	Duan et al. 2013
Graphene nanoparticles	High toxicity due to the graphene adherence to the surface of the chorion, causing hypoxia and hatching delay	Ou et al. 2016
Polymeric nanoparticles with antioxidant activity	Not toxic	Vong, Kobayashi and Nagasaki 2016
Chitosan-coated PLGA mucoadhesive nanoparticles containing BENZ (NPMAs)	Not toxic	Charlie-Silva et al. 2020
Polyhydroxyalkanoates	Not toxic	Li et al. 2016
Hydrogel scaffolds based on alginate, gelatin, 2-hydroxyethyl methacrylate, and hydroxyapatite	Not toxic	Tomić et al. 2021

Using natural origin and/or synthetic polymers to design advanced nanomaterials has significant advantages (ease of chemical modification, fine tunability, controllable size/morphology, good stability) which make them favorable candidates for biomedical applications (Banik et al. 20016; Tomić et al. 2010; Babić et al. 2020). During the last decade, the zebrafish embryo model due to its sufficient sensitivity and specificity to distinguish variations in toxicity mediated by similar materials was established as a useful method to evaluate the biocompatibility of the polymeric nanomaterials. The *in vivo* toxicity between polymeric nanoparticles and zebrafish mostly depends on their physicochemical properties where the surface charge is recognized as a crucial factor (Jia et al. 2019). Bodewein et al. (2016) compared the differences in the toxicity of anionic/cationic polyamidoamine (PAMAM) dendrimers of generations G3.0, 3.5, 4.0, 4.5 and 5.0 and cationic polypropylenimine (PPI) dendrimers G3.0, 4.0 and 5.0 in zebrafish embryos for 96 h, and revealed that cationic dendrimers severely decreased the survival rate of zebrafish embryos and impaired their cardiovascular functions, in contrast, anionic PAMAM dendrimers showed little to no zebrafish embryotoxicity (Bodewein et al. 2016). With the expansion of the use of nanomaterials in biomedicine, a more comprehensive understanding of the safety of the nanomaterials is imperative for improving the risk assessment of these nanomaterials. Chitosan nanoparticles were prepared to evaluate its toxicity on zebrafish embryo (*Danio rerio*) and revealed that the malformation, mortality and hatching rates of zebrafish embryo showed dose-dependent effects in the groups treated with both normal chitosan particles and chitosan nanoparticles and also showed that the toxicity of chitosan nanoparticles to zebrafish embryo was lower than normal chitosan particles (Wang et al. 2016). The zebrafish embryotoxic effects were analyzed in a concentration-dependent approach of chitosan nanoparticles and their Tween 80 modified counterparts which are commonly used brain-targeted vehicles and demonstrated that both nanoparticles exposure induced apparent increased mortality rate, decreased hatching rate, and rising malformations dose-dependently. Among that, disturbed neurobehavioral activities, axonal growth of motor neurons and muscle development were also detected in zebrafish embryos and larvae exposed to both nanoparticles at the concentration of ½ LC_{50} (Yuan et al. 2016). Chitosan polymer coated with reduced graphene oxide and implanted with Fe_3O_4 nanoparticles was designed for cancer-targeted fluorescence imaging and targeted drug delivery and to assess the biocompatibility and *in vivo* toxicity of the nanocomposite zebrafish (*Danio rerio*) model was used and revealed that the different material combinations in the nanocomposite did not impact the survival rate of zebrafish at various developmental stages (from 0 to 72 h) and this finding was related to the utilization of chitosan by the embryo as nourishment (Karthika et al. 2020). The toxic mechanism of low-molecular-weight chitosan (LMWCS) using zebrafish (*Danio rerio*) liver (ZFL) cell line, zebrafish larvae, and adult fish were reported and revealed that LMWCS rapidly induced cytotoxicity of ZFL cells and death of zebrafish, LMWCS disrupt membrane of epithelial cells in zebrafish larvae which induce breakage of the yolk; adult fish exhibited hypoxia before death due to multiple damages induced by LMWCS (Chou et al. 2020). Monomethoxy(polyethylene glycol)-poly(DL-lactic-co-glycolic acid)-poly(L-lysine)(PEAL) were synthesized and examined biocompatibility of the PEAL and the impacts of LA and GA ratio (LA/GA=60/40, 70/30, 80/20) in the main material (He et al. 2014). Different concentrations of PEAL nanoparticles co-incubated with zebrafish embryos at 6-72 h post-fertilization did not induce significant effects only 1000 or 2000 μM of nanoparticles possess some effects. The obtained results represent the PEAL copolymer with excellent biocompatibility as a promising implant candidate (He et al. 2014).

It was applied sequential reversible addition-fragmentation chain transfer (RAFT) polymerization with subsequent removal of acetyl groups using tetra-butylammonium hydroxide (TBAOH) to obtain a series of galactose-based glycopolymers, poly(N-(prop-2-enoyl)-β-D-galactopyranosylamine)-b-poly (N-isopropyl acrylamide) (pGal(OH)-b-pNIPAA). Collaborative assembly via TBAOH treatment was applied to polymers loaded with hydrophilic doxorubicin hydrochloride drug (Li 2020). By this procedure, pGal-b-pNIPAA/doxorubicin (DOX) delivery nanoparticles (GND NPs) with uniform size, high drug loading efficiency and encapsulation were obtained. The collaborative assembly strategy leads to the well-controlled formation of particles in comparison with conventional adsorption methods. The GND NPs rapidly disintegrate in acidic media leading to fast and uncontrolled release of DOX. Cellular experiments showed that pGal-b-pNIPAA/DOX have high cellular uptake by HepG2 cells and anti-tumor efficacy in vitro. The transgenic zebrafish TO(KrasG12V) model in vivo proved that GND NPs III had more sustainable and effective anti-tumor effects than free DOX.

Chitosan nanoparticles (CNPs) were synthesized by the ionic gelation method using medium molecular weight chitosan (MMW-C) (Nikapitiya et al. 2018). CNPs had an average particle size and zeta potential of 181.2 nm and + 37.2 mv. The toxicity effect of CNPs and MMW-C to zebrafish was evaluated by delayed hatching, survival rate analysis and malformations of embryos upon treatment. The toxicity effect of CNPs in hatched larvae was minimal at 5 μg/ mL compared to MMW-C exposure. Immunomodulatory properties investigated in zebrafish larvae showed that in larvae exposed to CNPs 5 days post fertilization (5 dpf), immune-related (*il-1β*, *tnf-α*, *il-6*, *il-10*, *cxcl-18b*, *ccl34a.4*, *cxcl-8a*, *lyz-c*, *defβl-1*, *irf-1a*, *irf-3*, *MxA*) and stress (*hsp-70*) response genes were induced. At 72 hours after infection with pathogenic *Aeromonas hydrophila* zebrafish larvae (at 5 dpf stage) exposed to CNPs at 5 μg/mL showed a higher survival rate compared to controls. Besides, zebrafish exposure to CNPs at 5 μg/mL could enhance the immune responses and develop the disease resistance against *A. hydrophila*, due to the strong immune-modulatory properties of CNPs.

Liver targeted drug delivery system for cancer therapy was designed using pullulan stabilized gold nanoparticles (PAuNPs), following green chemistry principles, using gold nanoparticles and natural, biocompatible, liver cell-specific biopolymer pullulan (Ganeshkumar et al. 2014). Nanoparticles were coupled with 5-Fluorouracil (5-Fu) and folic acid (Fa) to be used as targeted drug delivery vehicles and for cancer imaging. The toxicity of the drug 5-Fu, 5-Fu adsorbed gold nanoparticles (5-Fu@AuNPs) and Fa-coupled 5-Fu adsorbed gold nanoparticles (5-Fu@AuNPs-Fa), was studied *in vivo* using zebrafish embryo model. The *in vitro* cytotoxicity studied against HepG2 cells revealed that the amount of 5-Fu required to achieve 50% of the growth of inhibition (Ic50) was much lower in 5-Fu@AuNP-Fa than in free 5-Fu and in 5-Fu@AuNP. The *in vivo* and *in vitro* cytotoxicity results indicate that 5-Fu@AuNPs-Fa *have* the necessary qualities to be used in drug delivery applications for *liver cancer-targeted* carrier systems.

Biocompatible nanoparticles based on an inherently fluorescent polypyrazoline (PPy) core and a poly(ethylene glycol) (PEG) shell, with characteristic size, photophysical properties, good colloidal and optical stabilities, and ability to avoid the reticuloendothelial system were synthesized (Mane et al. 2018). Photoinitiated step-growth polymerization was used to obtain PPy-PEG amphiphilic block copolymer. PPy nanoparticles (PPy NPs) were then produced by a direct nanoprecipitation procedure. PPy NPs exhibit a large Stokes shift ($\Delta\lambda$ = 160 nm or Δv > 7000 cm^{-1}) with visible light excitation and strong colloidal stability. PPy NPs exhibit good biocompatibility. After injection in zebrafish embryos, PPy NPs were practically entirely observed in the circulating bloodstream, with little decrease with time and no acute morphological and behavioral abnormalities in embryos treated with NPs were observed.

Imaging of fluorescent PEG-stabilized PPy nanoparticles (NPs) for *Mycobacterium tuberculosis* (Mtb) and the fish tuberculosis (TB) agent *Mycobacterium marinum* (Mm) treatment was presented (Fenaroli et al. 2014). It was used the zebrafish embryo model for parallel imaging of fluorescent PEG-stabilized Ppy nanoparticles (NPs) and the fish tuberculosis (TB) agent. In the study transgenic lines of macrophages, neutrophils, and endothelial cells expressing fluorescent markers were used to visually detect the effects of NPS on zebrafish embryo treatment. Phagocytosis by blood macrophages was accomplished by injection of Mm in the zebrafish venes.

The introduced Mm started to form aggregates of infected cells after 3 days of retention in the vasculature. The NPs which were subsequently injected got rapidly into both infected and uninfected macrophages. Efficient targeting of uninfected macrophages to the site of infection, ensuring efficient targeting in the granuloma. The drug (Rifampicin) loaded in the NPs provided better embryo survival and a lower concentration of bacteria.

Redox and pH dual-responsive nanoparticles (NPs) loaded with doxorubicin (DOX) were prepared for cancer therapy as it has been shown that NPs were very effective in cancer treatment (Ang et al. 2017). Nanoparticles based on poly(acrylic acid) (PAA) as the backbone, were used to prepare self-assembly of β-cyclodextrin (β-CD) containing PAA (PAA-CD), adamantane (AD) conjugated PAA bridged by a disulfide bond (PAA-SS-AD), and AD mono-terminated PEG_{5000} by the supramolecular approach. The release of DOX is enabled by disulfide bonds in the polymeric network of NPs in the cancer cells endoplasm, due to its reducing effect. The DOX loaded in NPs can be released in the acidic environment, as well. *In vitro* studies have shown an excellent effect of dual-responsive nanoparticles in inducing cell death and apoptosis. After 3 days of treatment, DOX-loaded nanoparticles were more effective in improving zebrafish larvae viability, compared with zebrafish larvae treated with free DOX.

Fluorescent squaramide-based supramolecular polymer nanoparticles were prepared to as favorable drug delivery nanocarriers (Talens et al. 2020). These nanoparticles were prepared in a range of morphologies from fibers to spheres by self-assembling bola amphiphiles in water to form aggregates of nanoscale dimensions. Different nanoparticle morphologies maintain their surface chemistry and practically neutral surface charge. In the co-assembly approach, a sulfo-cyanine-labelled monomer was applied to track nanoparticle *in vivo* biodistribution and clearance in optically transparent zebrafish embryos. Fibrillar morphologies displayed a low circulation time with rapid association with venous endothelial cells, whereas spherical nanoparticles demonstrated significantly greater mobility with longer circulation times and improved distribution over the zebrafish. By using transparent zebrafish a molecular view into the nanoparticle interaction with scavenger endothelial cells was enabled. The important role of nanocarrier shape and size in nanoparticle clearance has been established by SECs through the Stabilin-2 receptor. The zebrafish model enabled tracking *in vivo* the route of nanoparticles at the cellular level and helped establish their structure-activity relationships, which pave the way for new strategies for drug delivery to a particular target.

Polymer coated ZnO nanoparticles were produced to obtain lower toxicity of sun blockers and cosmetic products. Chitosan (CTS) and poly(ethylene glycol) (PEG) were used for coating the ZnO nanoparticles to produce (ZnO-CTS) and (ZnO-PEG) nanoparticles (Girigoswami et al. 2015). In that way, the UV blocking efficacy of ZnO nanoparticles due to photocatalysis was prevented and the stability and UV absorption efficacy of nanoparticles increased. ZnO-PEG showed better UV attenuation. The toxicity of polymer-coated ZnO-CTS, ZnO-PEG nanoparticles and ZnO nanoparticles without polymer was tested *in vivo* using the zebrafish embryo model. The zebrafish embryo survival was higher for ZnO-CTS nanoparticles than for the ZnO nanoparticles. Delayed hatching was observed for PEG-coated nanoparticles. The amount of nanoparticles deposited on the zebrafish eggs was lowest for ZnO-CTS and then increased for ZnO-PEG to be the highest for ZnO. The authors presumed that the lower deposition of ZnO-CTS nanoparticles on eggs enabled embryos easier breathing and higher survival.

Chitosan was cross-linked ionically with sodium tripolyphosphate to obtain cross-linked chitosan nanoparticles. Two types of such nanoparticles with different particle sizes were synthesized (Yuan et al. 2011) and the influence on zebrafish embryos was examined. The result of incubation of zebrafish embryos with chitosan nanoparticles, using ZnO nanoparticles as the positive control, revealed that the hatching rate decreased and mortality increased, depending on nanoparticle concentration. Chitosan nanoparticles 200 nm in size also led to malformations in zebrafish embryos, such as the bent spine, pericardial edema, and opaque yolk, as well as increased rate of cell death and a high expression of reactive oxygen species. A high expression of heat shock protein 70 pointed out that physiological stress in zebrafish was induced by chitosan nanoparticles.

An advanced method to promote expeditious semi-automated fluorescence method to quantify the cellular responses of zebrafish embryo to polystyrene and poly(ε-caprolactone) microspheres, using ImageJ and the plugin ObjectJ project file “Zebrafish-Immunotest” was developed (Zhang et al. 2017). Treating the zebrafish embryo with polystyrene and poly(ε-caprolactone) microspheres ordered inflammatory cell response, similar to that obtained for mammalian animal models, was observed. The change of microspheres size of 10 and 15 μm did not change inflammatory cell responses. Infiltration of neutrophils and macrophages was more pronounced in poly(ε-caprolactone) than for polystyrene microspheres, as a potential means for inhibiting selective aspects of inflammation. This research demonstrated how the zebrafish embryos model can be useful for *in vivo* assessment of inflammatory cell responses of individual biomaterials.

The zebrafish embryo model as a favorable animal model was presented to study the developmental toxicity of low generation polyamidoamine (PAMAM) dendrimers (Heiden et al. 2007). Dendrimers represent polymeric molecules having a highly ordered structure of radially symmetric branches, with a typically symmetric core, an inner shell, and an outer shell, of nano-size with monodisperse particle size distribution. The developmental toxicity was determined for low generation PAMAM dendrimers (G3.5 and G4), and the result was compared with RGD-conjugated forms using the zebrafish embryo model. For G4 dendrimers it was found that amino end groups lead to toxicity and reduce growth and development of zebrafish embryos for concentrations below lethal, while G3.5 dendrimers, with carboxylic acid groups, were not toxic to zebrafish embryos. Using RGD-conjugated G4 and G3.5 dendrimers. The toxic effect of RGD-conjugated dendrimers revealed decreased toxicity for G4 conjugates. G3.5 conjugates were not toxic even for the highest concentrations applied, showing that they can be used in drug delivery systems.

Nanotoxicological and teratogenic effects of PAMAM dendrimers DG4 and DG4.5 were studied in zebrafish embryos and larvae, to define biocompatibility and therapeutic efficacy of polyamidoamine (PAMAM) dendrimers using *in vivo* method (Calienni et al. 2017). The authors tried to establish extensive toxicity screening of DG4 and DG4.5 and examine how their surface charge affects the zebrafish embryos and larvae. The surface charge, determined in both larvae and embryos, revealed that the positive surface charge of DG4 was more toxic than the negative one of DG4.5. Teratogenic effects of DG4 and DG4.5 dendrimers were manifested in larvae, while DG4 led to lethal effects in both zebrafish embryos and larvae. The plan of action proposed in combination with results realized may be useful to trace the way for further characterization of toxicity of similar nanoparticles.

With the rapid development of engineered hydrogel scaffolding biomaterials as one of the most attractive polymeric biomaterials for regenerative engineering which can be designed into tissue mimetic scaffolds to support cell growth due to their similarity to the native extracellular matrix, *in vivo* tests for evaluating the biocompatibility and biological safety of biomaterials are in crucial require (Tomić et al. 2021; Babić et al. 2018). Zebrafish has long been considered as the "gold standard" for biocompatibility and biological safety assessments of materials and chemicals due to its high fecundity, cost-effectiveness, well-characterized developmental stages, and optical transparency (Williams et al. 2015; Jia et al. 2019). Accordingly, zebrafish hold great potential for high-throughput screening of biocompatibility and biological safety. Among that, the injectability of hydrogel scaffolds renders them ideal for testing in zebrafish. Hsieh and collaborators tested the potential of 3D bioprinted polyurethane containing neural stem cells based on hydrogel in neuronal regeneration in zebrafish embryos and adults (Hsieh et al. 2015). The brain of embryos was damaged before the injection of the hydrogel by rearing in 2% ethanol and for adult zebrafish brain damage was achieved using a syringe to cause mechanical disruption. Then, net hydrogel, a hydrogel with murine neural stem cells, and neuronal stem cells were injected into zebrafish. From the evaluation of the obtained results, behavior and mortality, it was revealed that murine neuronal stem cells-laden polyurethane scaffolds with a concentration of 25% polyurethane had excellent proliferation and differentiation and could rescue the function of the impaired nervous system (Hsieh et al. 2015). The zebrafish embryo model was used to evaluate chitosan based injectable, self-healing hydrogel for healing nerve system deficits (Tseng et al. 2015). In the zebrafish neural injury model, injection of the self-healing chitosan based hydrogel with neurosphere-like progenitors produced a remarkable healing effect on neural development (81% recovery). The obtaining data represent the injectable chitosan based hydrogels encapsulating aggregated neural progenitor as a promising strategy in repairing the central nervous system and for rescuising the neural function (Tseng et al. 2015). A novel degradable chitosan-fibrin based hydrogel was presented as an *in vivo* injectable hydrogel which induces an angiogenic response in the perivitelline space of zebrafish (Hsieh et al. 2017). Due to support angiogenesis which is in turn responsible for bone formation injected chitosan-fibrin hydrogel is a promising candidate for bone tissue engineering applications (Vimalraj et al. 2021). Chitosan-fibrin injection increased EGFP gene expression and vascular endothelial cell population compared to chitosan hydrogel or control. The magnificent self-healing and angiogenic capacities of the chitosan-fibrin hydrogel may be associated with the formation of an interpenetrating polymeric network structures between chitosan and fibrin which offers new possibilities for various biomedical applications.

Zebrafish Interaction with Alginate/Gelatin/(Meth)Acrylate Based Hydrogels and Hydrogel Scaffolds

Zebrafish Interaction with Poly(β-Amino Ester) Hydrogels Based on Diacrylate and Glycine

Biodegradable poly (*β*-amino ester) zwitterionic hydrogels, based on glycine and di(ethylene glycol) diacrylate were designed (Filipović et al. 2019). PBAE macromers were first synthesized using Michael addition (Scheme 2). The diacrylate/amine stoichiometric ratio was changed(maintaining it >1), macromers with different chemical structures, containing acrylate end-groups were obtained, in a modified procedure (Filipović et al. 2018). Amino acid glycine was used as the amine component of the PBAE macromer. The hydrogels were synthesized from macromers by free radical polymerization. Unlike traditional zwitterionic polymers, poly(*β*-amino esters) are completely biodegradable, due to the existence of biodegradable ester bonds in the polymer backbone. These new biomaterials were tested for biocompatibility and therapeutic efficacy. Reduction and replacement of the requirement to conduct an experiment using full-grown animals has been achieved through utilising the zebrafish embryo model, a promising bridge model between *in vitro* and *in vivo* researches.

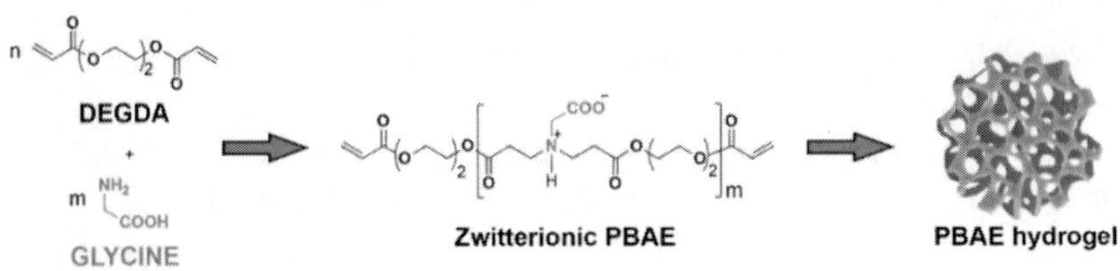

Scheme 2. Synthesis path of PBAE macromers and PBAE hydrogels formation.

Hydrogel material solution was obtained by extracting the finely ground hydrogel samples for 3 days at 37°C (180 rpm) to obtain the concentration of 200 μg cm^{-3}. The zebrafish embryos were exposed to these solutions for 5 days in the sixth stage of the hpf phase. The results of the study showed that the samples H1 and H2 exhibited no toxic effects at a concentration of 150 μg cm^{-3}, and sample H3 did not exhibit any toxic effects, even at the higher concentration. The treatment of zebrafish embryos with these hydrogel extracts did not cause death, development of abnormalities or any cardiovascular disturbance (change of heart morphology, appearance of pericardial edema or reduced heart rate) or signs of hepatotoxicity (liver necrosis or lower yolk resorption).

At the highest concentration (200 μg cm^{-3}) hydrogel samples H1 and H2 displayed low hepatotoxicity, which was exhibited as somewhat darker liver and lesser resorption of the yolk, compared to the zebrafish embryos that were not exposed to the extract solution (Figure 2). Hepatotoxic effects were not present with lower concentrations of the material extract. As these effects were not exhibited when exposing the zebrafish embryos to material H3, which contains lower amounts of glycine, compared to samples H1 and H2, the increased hepatotoxicity can be a result of higher concentrations of glycine remains left after the degradation process. The PBAE hydrogels with glycine, as well as their degradation products, contain nitrogen, which can be metabolized into ammonia and induce local damage in the zebrafish embryo. The hydrogel sample with the lowest amount of nitrogen does not exhibit any hepatotoxicity, even when the highest concentration of the hydrogel extract was used. The results obtained in the *in vivo* experiments confirmed the results from the *in vitro* probes since the examined hydrogel extracts did not have any damaging effects on the surface of skin and fins of zebrafish, which consist mostly of fibroblast, making them good candidates for testing in *in vivo* wound healing therapy.

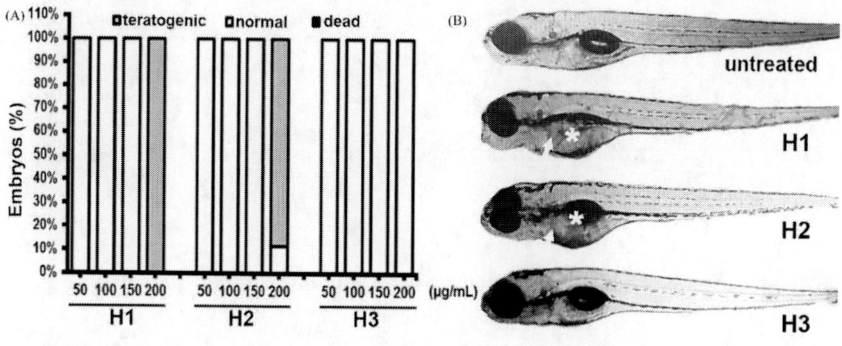

Figure 2. Survival, teratogenicity and morphology of 120 hpf-old zebrafish embryos exposed to the ground hydrogels (200 μg/mL); Embryos exposed to H3 developed without signs of toxicity, the embryos upon H1 and H2 showed dark liver (arrow) and weakly resorbed yolk (asterisk), as the signs of hepatotoxicity.

Zebrafish Interactions with 2-Hydroxyethyl Methacrylate/Gelatin Hydrogel Scaffolds

Two series of porous and degradable hydrogel scaffolds were designed using natural origin polymers-alginate and gelatin, and synthetic monomeric component 2-hydroxyethyl methacrylate (tags AGH), by cryogelation and porogengelation (Scheme 3).

We have studied the impact of composition and method of preparation on biocompatible properties of the obtained scaffolds. All samples were characterized *in vivo* (zebrafish *Danio rerio*) biocompatible properties. The obtained results show that cryo samples, especially with gelatin show favorable biocompatible properties in comparison with PHEMA samples, which makes these materials highly attractive for biomedical uses.

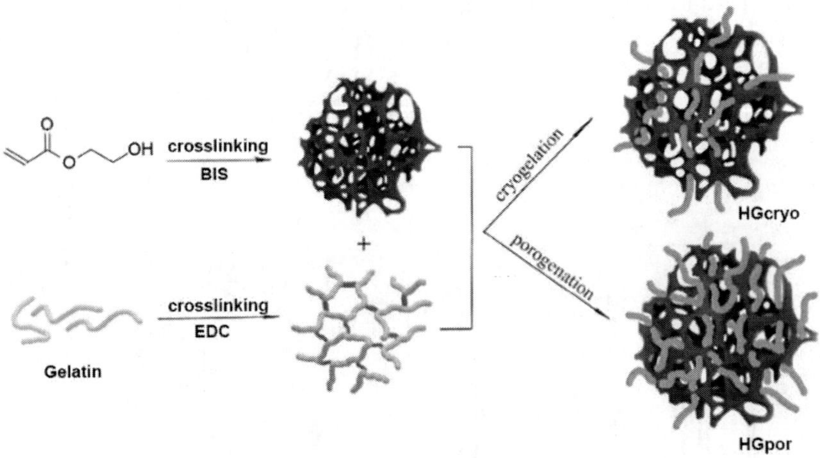

Scheme 3. Synthesis path of cryo and porogen hydrogel scaffolds based on 2-hydroxyethyl methacrylate and gelatin formation.

Biocompatibility assay of the materials has been tested *in vivo* using the zebrafish model for five days, using indirect assay (material extracts) as well as direct assay (ground materials). Obtained data showed that none of the tested materials caused lethal or teratogenic effects in the treated embryos during the 96-h treatment (Figure 3). Therefore, it can be said that regardless of the methodology of preparation or the addition of gelatin, materials prepared in this research were biocompatible and safe for the living organism under tested conditions (Tomić et al. 2020).

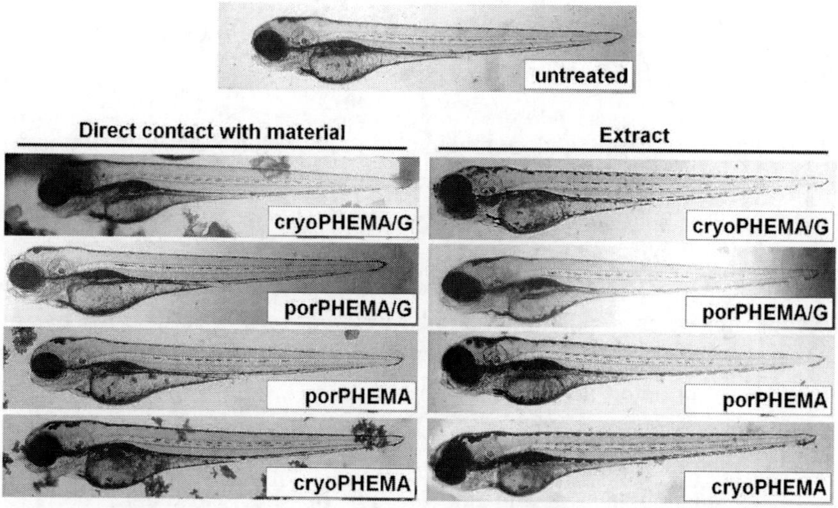

Figure 3. *In vivo* interactions of zebrafish embryos with PHEMA/G and PHEMA hydrogel scaffolds at 96 hpf. Wild type zebrafish were grown in the presence of the ground PHEMA/G scaffolds (left) and the material extract (100% v/v) (right).

To evaluate whether synthesized PHEMA/G scaffolds have the potential to induce inflammation or the immunosuppressive effect, the experiments were performed on transgenic zebrafish *Tg*(*mpx*: EGFP) embryos whose neutrophils expressing EGFP enable direct visualization upon applied treatments. The data showed that none of the tested materials provoked tissue inflammation which should be manifested as an increased neutrophil number when the embryos developed in direct contact with tested materials. However, the embryos developed in the material extract of porPHEMA (100% concentration) had slightly reduced neutrophils number (by 10%), while those in close contact with this material showed an even lower number of neutrophils (by 22%), compared to the control group indicating suppression of the immune response (Tomić et al. 2020).

Taken together these data indicate that tested materials could be regarded as non-toxic and with a high potential to be used in human use, even in the case of porPHEMA which induced a weak neutrophils-suppression effect as that could be beneficial in the aspect of suppression of inflammation.

Zebrafish Interactions with 2-Hydroxyethyl Methacrylate/ Gelatin/Alginate Hydrogel Scaffolds

Two series of porous and degradable hydrogel scaffolds were synthesized using natural origin polymers-alginate and gelatin, and synthetic monomeric component 2-hydroxyethyl methacrylate (tags HGA), using cryogelation and porogengelation (Scheme 4).

Hydrogel scaffolds were designed in three steps-first step of preparation, porous and mechanically strong polymeric network was obtained by free-radical crosslinking polymerization of 2-hydroxyethyl methacrylate in the presence of initiator, activator, and crosslinking agent. In the second step, an additional polymeric network was created by the crosslinking of gelatin. The third step-alginate chains are loaded to intertwine in the networks.

The main focus during the design of the scaffolding polymeric biomaterials to be used in regenerative medicine is the first step related to testing the biocompatibility of the matrix in an in vitro and in vivo system. The biocompatibility of HGA hydrogel scaffolds was examined in a cytotoxicity test on normal human fibroblasts (MRC5). All materials, when supplied directly to cells was shown to support the accumulation of cells on its surface. The obtained values confirm that HGA hydrogel scaffolds show a high level of cytocompatibility, and therefore suitable for applications as biomaterials in tissue regeneration.

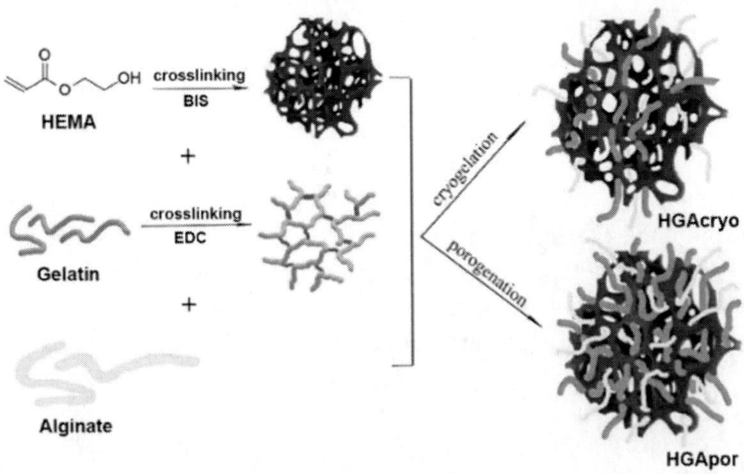

Scheme 4. Synthesis path of cryo and porogen hydrogel scaffolds composed of alginate, gelatin, and 2-hydroxyethyl methacrylate formation.

To detect whether the alginate/gelatin/methacrylate based materials could be used for medical applications, toxicity *in vivo* were performed using the zebrafish model. The zebrafish model was accepted as a valid alternative to mammalian models (rats mice) for toxicity and biocompatibility evaluation of novel biomaterials, nanomaterials, and drug carriers owing to the high molecular, genetic, physiological, and immunological similarity to humans, and good correlation in response to pharmaceuticals and bioactive compounds (Lieschke and Currie 2007; MacRae and Peterson 2015), simplifying thus the path to clinical trials and reducing the failure at later stages of testing (Barros et al. 2008; MacRae and Peterson 2015).

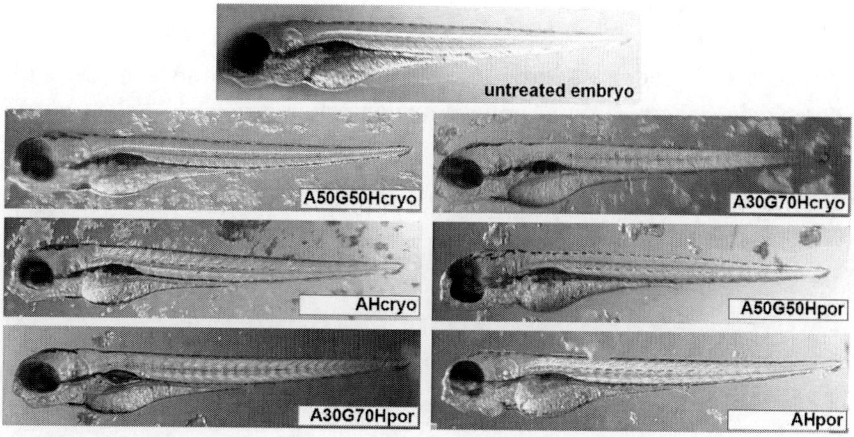

Figure 4. Morphology of zebrafish embryos in contact with HGA hydrogel scaffolds.

Given that the organisms at embryonic developmental stages are more sensitive to the chemical insults than as an adult, we exposed zebrafish embryos at the 6 hpf stage indirectly, to the material's extracts (embryo water in which the ground materials were extracted over 72 h at 37°C, for 5 days. Obtained data showed that the AGHcryo and AGHpor scaffolds tested, regardless of the method was no toxic to the zebrafish embryos, since there was no lethality, developmental malformation, cardiovascular disorders, or signs of hepatotoxicity (the liver necrosis and the yolk retention) (Figure 4), indicating the safety of the tested materials. Recently, we demonstrated the safety and biocompatibility (no inflammation and no immunosuppression) of 2-hydroxyethyl methacrylate based hydrogels using the zebrafish model (Tomić et al. 2020). In this study, we showed that AGH hydrogel scaffolds present also nontoxic materials, indicating their favorable potential in biomedical applications. Otherwise, the alginate based materials have been applied as safe biomaterials for the food packing and oral delivery to zebrafish larvae or adults (Onal and Langdon 2000; Lin et al. 2016) and more recently as the carriers of drugs to improve their therapeutic outcome (controlled release, reduced toxicity, increase in efficacy) (Tzankova et al. 2017; Gao et al. 2017).

Zebrafish Interactions with 2-Hydroxyethyl Methacrylate/ Gelatin/Alginate/Apatite Hydrogel Scaffolds

Porous and degradable hydrogel scaffolds were successfully synthesized by combining natural origin polymers-alginate and gelatin, and synthetic monomeric component 2-hydroxyethyl methacrylate (tags AGH) loaded with inorganic agent-hydroxyapatite (tag AGH/HAp) (Scheme 5).

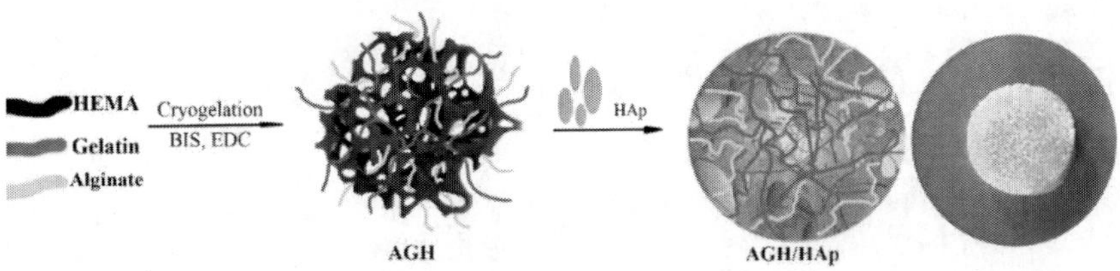

Scheme 5. Schematic representation of the polymeric hydrogel scaffolds creation composed of alginate, gelatin, 2-hydroxyethyl methacrylate and hydroxyapatite.

Biocompatibility *in vivo* of alginate, gelatin, 2-hydroxyethyl methacrylate and apatite based biomaterials we evaluated using the zebrafish model. The zebrafish model was accepted as a valid alternative to mammalian models (rats mice) for toxicity and biocompatibility evaluation of novel biomaterials owing to the high molecular, genetic, physiological, and immunological similarity to humans, and good correlation in response to pharmaceuticals and bioactive compounds (Lieschke, G.J. and Currie 2007; MacRae and Peterson 2015) thus simplifying the path to clinical trials and reducing the failure at later stages of testing (Barros et al. 2008).

Given that the organisms at embryonic developmental stages are more sensitive to the chemical insults than as an adult, we exposed zebrafish embryos at the 6 hpf stage indirectly, to the material's extracts (embryo water in which the ground materials were extracted over 72 h at 37°C, for 5 days.

The obtained data showed that the tested AGH and AGH/HAp scaffolding materials were not toxic to zebrafish embryos, because there were no lethality, developmental malformations, cardiovascular disorders or signs of hepatotoxicity (liver necrosis and yolk retention) (Figure 5) (Tomić et al. 2021). It can be said that the synthesized materials are acceptable for biomedical applications. It has been shown following previous research that polymeric biomaterials containing natural polymers can be applied as safe biomaterials for the food packaging and oral delivery to zebrafish larvae or adults (Onal, U. and Langdon 2000; Lin et al. 2016) and more recently as the carriers of therapeutically active agents to improve their therapeutic outcome (controlled release, reduced toxicity, increase in efficacy) (Tzankova et al. 2017; Gao, et al. 2017).

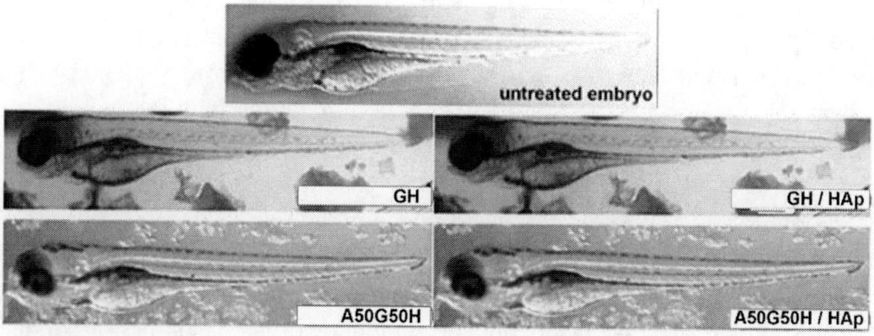

Figure 5. Morphology of zebrafish embryos exposed to GH, GH/HAp, A50G50H and A50G50H/HAp hydrogel scaffolds.

Conclusion

Biomaterials possess an expansive scope of applications, for example, clinical implants, including heart valves, stents, and grafts; artificial joints, tendons, and ligaments; hearing loss scaffolds; dental materials; and devices that invigorate nerves. Due to the high demand for the development of new biomaterials, a rapid and reliable testing platform is required. At present, biomaterials are assessed *in vitro* examinations, which are misrepresented, and implantation in rat models, which are expensive, tedious, and to a great extent depend on fixed histological endpoints. Alternatively, the zebrafish model that grants *in vivo* imaging and similarly straightforward hereditary control to the investigation of human disease. A continuous examination is absurd with mammalian models. Zebrafish give particular focal points over other *in vivo* models because there is a wide library of available transgenic fluorescent lines and they are reasonable, optically clear, easy gene edition, highly reproducible, and permit whole-animal live imaging investigation. Zebrafish have been utilized widely as an animal model to treat many human diseases, which are getting progressively significant in the design and analysis of new biomaterial therapies.

Thanks to zebrafish models, in recent years, investigations in the field of human diseases faced remarkable progressions that revealed the importance of its uses in different studies. Recent investigations have been implemented in zebrafish in endocrine-related disorders using "omics" technologies. In fact, due to the specific characteristics of zebrafish in providing the possibility of rapid screening with the ability to distinguish molecular changes based on sex and organ specification, it can be considered as a suitable choice in this regard. Moreover, since zebrafish can be utilized for high-throughput non-biased chemical screening, there is a growing rate in its employing for *in vivo* drug discovery researches. Interestingly, its notable similarity to humans and providing the accessibility to a wide set of genetic tools for gene mutation, in addition to its rapid development and limited ethical issues introduces zebrafish as a potential choice for individualized drug therapy studies. The outlook of using zebrafish in personalized regenerative medicine seems promising due to the same advantages which zebrafish has in cancer studies. Furthermore, advances in visualization methods, gene targeting, and chemical screening in zebrafish models can improve future perspectives and findings.

The novel biocompatible, porous, and degradable hydrogels based on amino acid and acrylates and hydrogel scaffold based on alginate, gelatin, 2-hydroxyethyl methacrylate, and with inorganic agent hydroxyapatite were successfully designed. *In vivo* zebrafish model study confirmed that all biomaterials possess favorable biocompatibility. All these unique properties indicate that designed hydrogels and hydrogel scaffolds are promising candidates for regenerative engineering biomaterials.

Acknowledgments

This research was funded by the SCOPES program of the Swiss National Science Foundation (SNSF) and the Swiss Agency for Development and Cooperation (SDC) (Grants No IZ73ZO_152327). This research has been supported by the Ministry for Education, Science and Technological

Development of the Republic of Serbia (Contracts No. 451-03-9/2021-14/172062 and 451-03-9/2021-14/172026).

Conflict of Interest

The authors declare that there is no conflict of interest regarding the publication of this paper.

References

Ali, Z., Islam A., Sherrell P., Le-Moine M., Lolas G., Syrigos K., Rafat M., Jensen L. D. (2018). Adjustable delivery of pro-angiogenic FGF-2 by alginate: collagen Microspheres. *Biology Open*, 7 bio027060. doi:10.1242/bio.027060.

Amsterdam, A. and Hopkins, N. (2006). Mutagenesis strategies in zebrafish for identifying genes involved in development and disease. *Trends in Genetics*, 473-478.

Ang, C. Y., Tan, S. Y., Teh, C., Lee, J. M., Wong, M. F., Qu, Q., Poh, L. Q., Li, M., Zhang, Y., Korzh, V. and Zhao, Y. (2017). Redox and pH Dual Responsive Polymer Based Nanoparticles for In Vivo Drug Delivery. *Small*, 1602379.

Arjmand, B., Tayanloo-Beik, A., Foroughi Heravani, N., Alaei, S., Payab, M., Alavi-Moghadam, S., Goodarzi, P., Gholami, M. and Larijani, B. (2020). Zebrafish for Personalized Regenerative Medicine; A More Predictive Humanized Model of Endocrine Disease. *Frontiers in Endocrinology,* 11. doi:10.3389/fendo.2020.00396

Azman, I., Mutalib, S. A., Yusoff, S. F. M., Fazry, S., Noordin, A., Kumaran, M. and Mat Lazim, A. (2015). Novel *Dioscorea hispida* starch-based hydrogels and their beneficial use as disinfectants. *Journal of Bioactive and Compatible Polymers*, 31(1), 42-59.

Babić, M. M., Božić, B. B., Božić, B. B., Ušćumlić, G. and Tomić, S. Lj. (2018). The innovative combined microwave-assisted and photo-polymerization technique for synthesis of the novel degradable hydroxyethyl (meth)acrylate/gelatin based scaffolds. *Materials Letters*, 236-240.

Babić, M. M., Vukomanović, M., Stefanič, M., Nikodinović-Runić, J. and Tomić, S. Lj. (2020). Controlled Curcumin release from hydrogel scaffold platform based on 2-hydroxyethyl methacrylate/gelatin/alginate/Iron(III) Oxide, *Macromolecular Chemistry and Physics*, 2000186.

Balitaan, J. N. I., Hsiao, C.-D., Yeh, J.-M. and Santiago, K. S. (2020) Innovation inspired by nature: Biocompatible self-healing injectable hydrogels based on modified-β-chitin for wound healing. *International Journal of Biological Macromolecules*, 723-736.

Banik, B. L., Fattahi, P. and Brown, J. L. (2016). Polymeric nanoparticles: the future of nanomedicine. *Wiley Interdisciplinary Reviews-Nanomedicine and Nanobiotechnology*, 271-299.

Barros, T. P., Alderton, W. K., Reynolds, H. M., Roach, A. G. and Berghmans, S. (2008). Zebrafish: an emerging technology for in vivo pharmacological assessment to identify potential safety liabilities in early drug discovery. *British Journal of Pharmacology*, 1400-1413.

Bodewein, L., Schmelter, F., Di Fiore, S., Hollert, H., Fischer, R. and Fenske, M. (2016). Differences in toxicity of anionic and cationic PAMAM and PPI dendrimers in zebrafish embryos and cancer cell lines. *Toxicology and Applied Pharmacology*, 83-92.

Brundo, M. V., Pecoraro, R., Marino, F., Salvaggio, A., Tibullo, D., Saccone, S., Bramanti, V., Buccheri, M. A., Impellizzeri, G., Scuderi, V., Zimbone, M. and Privitera, V. (2016). Toxicity Evaluation of New Engineered Nanomaterials in Zebrafish. *Frontiers in Physiology*, 7(130).

Buschmann, J. (2013). The OECD guidelines for the testing of chemicals and pesticides. *Methods in molecular biology*, 37-56.

Cairns, G., Burté, F., Price, R., O'Connor, E., Toms, M., Mishra, R., Moosajee, M., Pyle, A., Sayer, J. A. and Yu-Wai-Man, P. (2021). A mutant wfs1 zebrafish model of Wolfram syndrome manifesting visual dysfunction and developmental delay. *Scientific Reports*, 20491.

Calienni, M. N., Feas, D. A., Igartúa, D. E., Chiaramoni, N. S., Alonso, S. D. V. and Prieto, M. J. (2017). Nanotoxicological and teratogenic effects: A linkage between dendrimer surface charge and zebrafish developmental stages. *Toxicology and applied pharmacology*, 1-11.

Chalin, D., Bureau, C., Parmeggiani, A., Rochal, S., Kissa, K. and Golushko, I. (2021). Modeling and live imaging of mechanical instabilities in the zebrafish aorta during hematopoiesis. *Scientific Reports*, 9316.

Charlie-Silva, I., Feitosa, N. M., Gomes, J. M. M., Hoyos, D. C. d. M., Mattioli, C. C., Eto, S. F., Fernandes, D. C., de Andrade Belo, M. A. and Fraceto, L. F. (2020). Potential of mucoadhesive nanocapsules in drug release and toxicology in zebrafish. *PLOS ONE*, e0238823.

Cheng, T. Y., Chen, M. H., Chang, W. H., Huang, M. Y. and Wang, T. W. (2013). Neural stem cells encapsulated in a functionalized self-assembling peptide hydrogel for brain tissue engineering. *Biomaterials*, 2005-2016.

Chou, C. M., Mi, F. L., Horng, J. L., Lin, L. Y., Tsai, M. L., Liu, C. L., Lu, K. Y., Chu, C. Y., Chen, Y. T., Lee., Y. L. A. and Cheng, C. H. (2020). Characterization and toxicology evaluation of low molecular weight chitosan on zebrafish. *Carbohydrate Polymers*, 116164.

Costa, B., Estrada, M. F., Mendes, R. V. and Fior, R. (2020). Zebrafish Avatars towards Personalized Medicine—A Comparative Review between Avatar Models. *Cells*, 293.

Crecente-Campo, J., Guerra-Varela, J., Peleteiro, M., Gutiérrez-Lovera, C., Fernández-Mariño, I., Diéguez-Docampo, A., González-Fernández, Á., Sánchez, L. and Alonso, M. J. (2019). The size and composition of polymeric nanocapsules dictate their interaction with macrophages and biodistribution in zebrafish. *Journal of Controled Release*, 98-108.

d'Amora, M., Raffa, V., De Angelis, F. and Tantussi, F. (2021). Toxicological Profile of Plasmonic Nanoparticles in Zebrafish Model. *International Journal of Molecular Sciences*, 6372.

Dahlman, J. E., Barnes, C., Khan, O. F., Thiriot, A., Jhunjunwala, S., Shaw, T. E., Xing, Y. P., Sager, H. B., Sahay, G., Speciner, L., Bader, A., Bogorad, R. L., Yin, H., Racie, T., Dong, Y. Z., Jiang, S., Seedorf, D., Dave, A., Sandhu, K. S., Webber, M.J., Novobrantseva, T., Ruda, V. M., Lytton-Jean, A. K. R., Levins, C. G., Kalish, B., Mudge, D. K., Perez, M., Abezgauz, L., Dutta, P., Smith, L., Charisse, K., Kieran, M. W., Fitzgerald, K., Nahrendorf, M., Danino, D., Tuder, R. M., von Andrian, U. H., Akinc, A., Panigrahy, D., Schroeder, A., Koteliansky, V., Langer, R. and Anderson, D. G.. (2014). In vivo endothelial siRNA delivery using polymeric nanoparticles with low molecular weight. *Nature Nanotechnology*, 648-655.

Duan, J., Yu, Y., Shi, H., Tian, L., Guo, C., Huang, P., Zhou, X., Peng, S. and Sun, Z. (2013). Toxic effects of silica nanoparticles on zebrafish embryos and larvae. *PLOS ONE*, e74606.

Ducharme, N. A., Reif, D. M., Gustafsson, J. A. and Bondesson, M. (2015). Comparison of toxicity values across zebrafish early life stages and mammalian studies: Implications for chemical testing. *Reproductive Toxicology*, 3-10.

Faillaci, F., Milosa, F., Critelli, R. M., Turola, E., Schepis, F. and Villa, E. (2018). Obese zebrafish: A small fish for a major human health condition. *Animal Models and Experimental Medicine*, 255-265.

Fenaroli, F., Westmoreland, D., Benjaminsen, J., Kolstad, T., Skjeldal, F. M., Meijer, A. H., van der Vaart, M., Ulanova, L., Roos, N., Nyström, B., Hildahl, J. and Griffiths, G. (2014). Nanoparticles as drug delivery system against tuberculosis in zebrafish embryos: direct visualization and treatment. *ACS Nano*, 7014-7026.

Filipović, V. V., Babić, M. M., Godevac, D., Pavić, A., Nikodinović-Runić, J. and Tomić, S. Lj. (2019). *In Vitro* and *In Vivo* Biocompatibility of Novel Zwitterionic Poly(Beta Amino)Ester Hydrogels Based on Diacrylate and Glycine for Site-Specific Controlled Drug Release. *Macromolecular Chemistry and Physics*, 1900188.

Filipović, V. V., Božić Nedeljković, B., Vukomanović, M. and Tomić, S. Lj. (2018). Biocompatible and degradable scaffolds based on 2-hydroxyethyl methacrylate, gelatin and poly(β-amino ester) crosslinkers. *Polymer Testing*, 68.

Ganeshkumar, M., Ponrasu, T., Raja, M. D., Subamekala, M. K. and Suguna, L. (2014). Green synthesis of pullulan stabilized gold nanoparticles for cancer targeted drug delivery. *Spectrochimica Acta Part A: Molecular and Biomolecular Spectroscopy*, 64-71.

Gao, C., Tang, F., Gong, G., Zhang, J., Hoi, M. P. M., Lee, S. M. Y. and Wang, R. (2017). pH-Responsive prodrug nanoparticles based on a sodium alginate derivative for selective co-release of doxorubicin and curcumin into tumor cells. *Nanoscale*, 12533-12542.

Gao, Y. (2018). Design and application of dextran based cross-linked networks. *Doctoral Thesis, Leiden Institute of Chemistry (LIC)*, Faculty of Science, Leiden University, 63-79.

George, S., Xia, T., Rallo, R., Zhao, Y., Ji, Z., Lin, S. J., Wang, X., Zhang, H. Y., France, B., Schoenfeld, D., Damoiseaux, R., Liu, R., Lin, S., Bradley, K. A., Cohen, Y. and

Nel, A. E. (2011). Use of a High-Throughput Screening Approach Coupled with In Vivo Zebrafish Embryo Screening To Develop Hazard Ranking for Engineered Nanomaterials. *ACS Nano*, 1805-1817.

Girigoswami, K., Viswanathan, M., Murugesan, R. and Girigoswami, A. (2015). Studies on polymer-coated zinc oxide nanoparticles: UV-blocking efficacy and in vivo toxicity. *Materials Science and Engineering: C*, 501-510.

Hamilton, F. (1822). An account of the fishes found in the river Ganges and its branches. *Archibald Constable*.

He, Z., Wang, Q., Sun, Y., Shen, M., Zhu, M., Gu, M., Wang Y. and Duan, Y. (2014). The biocompatibility evaluation of mPEG-PLGA-PLL copolymer and different LA/GA ratio effects for biocompatibility. *Journal of Biomaterials Science, Polymer Edition*, 943-964.

Heiden, T. C., Dengler, E., Kao, W. J., Heideman, W. and Peterson, R. E. (2007). Developmental toxicity of low generation PAMAM dendrimers in zebrafish. *Toxicology and Applied Pharmacology*, 70-79.

Høgset, H., Horgan, C. C., Armstrong, J. P. K., Bergholt, M. S., Torraca, V., Chen, Q., Keane, T. J., Bugeon, L., Dallman, M. J., Mostowy, S. and Stevens, M. M. (2020). In vivo biomolecular imaging of zebrafish embryos using confocal Raman spectroscopy. *Nature Communications*, 6172.

Holler, K., Neuschulz, A., Drewe-Boß, P., Mintcheva, J., Spanjaard, B., Arsiè, R., Ohler, U., Landthaler, M. and Junker, J. P. (2021). Spatio-temporal mRNA tracking in the early zebrafish embryo. *Nature Communications*, 3358.

Hörstgen-Schwark, G. (1993). Production of homozygous diploid zebrafish (*Brachydanio rerio*). *Aquaculture*, 25-37.

Houart, C. (2001). Zebrafish as an Experimental Organism. In *eLS*.

Howe, K., Clark, M. D., Torroja, C. F., Torrance, J., Berthelot, C., Muffato, M. and Stemple, D. L. (2013). The zebrafish reference genome sequence and its relationship to the human genome. *Nature*, 498-503.

Hsieh, F. Y., Lin, H. H. and Hsu, S. H. (2015). 3D bioprinting of neural stem cell-laden thermoresponsive biodegradable polyurethane hydrogel and potential in central nervous system repair. *Biomaterials*, 48-57.

Hsieh, F. Y., Tao, L., Wei, Y., and Hsu, S. H. (2017). A novel biodegradable self-healing hydrogel to induce blood capillary formation. *NPG Asia Materials*, e363-e363.

Hu, Y. L., Qi, W., Han, F., Shao, J. Z. and Gao, J. Q. (2011). Toxicity evaluation of biodegradable chitosan nanoparticles using a zebrafish embryo model. *International Journal of Nanomedicine*, 3351-3359.

Jia, H. R., Zhu, Y.-X., Duan, Q. Y., Chen, Z. and Wu, F. G. (2019). Nanomaterials meet zebrafish: Toxicity evaluation and drug delivery applications. *Journal of Controlled Release*, 301-318.

Johnson, T. D. and Christman, K. L. (2013). Injectable hydrogel therapies and their delivery strategies for treating myocardial infarction. *Expert Opinion on Drug Delivery*, 59-72.

Kabashi, E., Brustein, E., Champagne, N. and Drapeau, P. (2011). Zebrafish models for the functional genomics of neurogenetic disorders. *Biochimica et Biophysica Acta*, 335-345.

Karthika, V., AlSalhi, M. S., Devanesan, S., Gopinath, K., Arumugam, A. and Govindarajan, M. (2020). Chitosan overlaid Fe(3)O(4)/rGO nanocomposite for targeted drug delivery, imaging, and biomedical applications. *Scientific Reports*, 18912.

Kimmel, C. B., Ballard, W. W., Kimmel, S. R., Ullmann, B. and Schilling, T. F. (1995). Stages of embryonic development of the zebrafish. *Developmental Dynamics*, 253-310.

Kürekçi, K. G., Mangit, K. E., Koyunlar, C., Unsal, S., Saglam, B., Ergin, B., Gizer, M., Uyanik, I., Düz, N. B., Korkusuz, P., Talim, B., Purali, N., Hughes, S. M. and Dincer, P. R. (2021). Knockout of zebrafish desmin genes does not cause skeletal muscle degeneration but alters calcium flux. *Scientific reports*, 7505-7505.

Lawson, N. D. and Weinstein, B. M. (2002). In vivo imaging of embryonic vascular development using transgenic zebrafish. *Developmental Biology*, 307-318.

Lee, S. L., Rouhi, P., Dahl Jensen, L., Zhang, D., Ji, H., Hauptmann, G., Ingham, P. and Cao, Y. (2009). Hypoxia-induced pathological angiogenesis mediates tumor cell dissemination, invasion, and metastasis in a zebrafish tumor model. *Proceedings of the National Academy of Sciences of the United States of America*, 19485-19490.

Leyk, J., Daly, C., Janssen-Bienhold, U., Kennedy, B. N. and Richter-Landsberg, C. (2017). HDAC6 inhibition by tubastatin A is protective against oxidative stress in a photoreceptor cell line and restores visual function in a zebrafish model of inherited blindness. *Cell Death & Disease*, e3028-e3028.

Li, J., Zhang, Y., Cai, C., Rong, X., Shao, M., Li, J., Yang, C. and Yu, G. (2020). Collaborative assembly of doxorubicin and galactosyl diblock glycopolymers for targeted drug delivery of hepatocellular carcinoma. *Biomaterials Science*, 189-200.

Li, X., Li, X., Chen, D., Guo, J. L., Feng, D. F., Sun, M. Z., Lu, Y., Chen, D. Y., Zhao, X. and Feng, X. Z. (2016). Evaluating the biological impact of polyhydroxyalkanoates (PHAs) on developmental and exploratory profile of zebrafish larvae. *RSC Advances*, 37018-37030.

Lieschke, G. J. and Currie, P. D. (2007). Animal models of human disease: zebrafish swim into view. *Nature Reviews Genetics*, 353-367.

Lin, L. J., Chiang, C. J., Chao, Y. P., Wang, S. D., Chiou, Y. T., Wang, H. Y. and Kao, S. T. (2016). Development of Alginate Microspheres Containing Chuanxiong for Oral Administration to Adult Zebrafish. *BioMed Research International*, 4013071.

Lin, M. Z., Li, W. X., Hu, T., Bu, H., Li, Z. L., Wu, T., Wu, X. X., Sun, C., Li, Y. And Jiang, G. B. (2021). One-step removal of harmful algal blooms by dual-functional flocculant based on self-branched chitosan integrated with flotation function. *Carbohydrate Polymers*, 117710.

Liu, Y., Hsu, Y. H., Huang, A. P. H. and Hsu, S. H. (2020). Semi-Interpenetrating Polymer Network of Hyaluronan and Chitosan Self-Healing Hydrogels for Central Nervous System Repair. *ACS Applied Materials & Interfaces*, 40108–40120.

Lu, W. N., Lü, S. H., Wang, H. B., Li, D. X., Duan, C. M., Liu, Z. Q., Hao, T., He, W. J., Xu, B., Fu, Q., Song, Y. C., Xie, X. H. and Wang, C. Y. (2009). Functional improvement of infarcted heart by co-injection of embryonic stem cells with temperature-responsive chitosan hydrogel. *Tissue engineering Part A*, 1437-1447.

Macdonald, N. P., Zhu, F., Hall, C. J., Reboud, J., Crosier, P. S., Patton, E. E., Wlodkowic, D. and Cooper, J. M. (2016). Assessment of biocompatibility of 3D printed photopolymers using zebrafish embryo toxicity assays. *Lab on a Chip*, 291-297.

MacGregor, J. T., Collins, J. M., Sugiyama, Y., Tyson, C. A., Dean, J., Smith, L., Andersen, M., Curren, R. D., Houston, J. B., Kadlubar, F. F., Kedderis, G. L., Krishnan, K., Li, A. P., Parchment, R. E., Thummel, K., Tomaszewski, J. E., Ulrich, R., Vickers, A. E. and Wrighton, S. A. (2001). In vitro human tissue models in risk assessment: report of a consensus-building workshop. *Toxicological Sciences*, 17-36.

MacRae, C.A. and Peterson, R.T. (2015). Zebrafish as tools for drug discovery. *Nature reviews Drug discovery*, 21-731.

Mane, S. R., Hsiao, I. L., Takamiya, M., Le, D., Straehle, U., Barner-Kowollik, C., Weiss, C. and Delaittre, G. (2018). Intrinsically Fluorescent, Stealth Polypyrazoline Nanoparticles with Large Stokes Shift for In Vivo Imaging. *Small*, 1801571.

McNamara, K., Tofail, S. A. M. (2017) Nanoparticles in biomedical applications. *Advances in Physics: X* 54-88.

Nikapitiya, C., Dananjaya, S. H. S., De Silva, B. C. J., Heo, G. J., Oh, C., De Zoysa, M. and Lee, J. (2018). Chitosan nanoparticles: A positive immune response modulator as display in zebrafish larvae against *Aeromonas hydrophila* infection. *Fish Shellfish Immunology*, 240-246.

OECD, T.N., 236: Fish Embryo Acute Toxicity (FET) Test. *OECD Guidelines for the Testing of Chemicals*, Section, 2013. 2.

Önal, U. and Langdon, C. (2000). Characterization of two microparticle types for delivery of food to altricial fish larvae. *Aquaculture Nutrition*, 6.

Ou, L., Song, B., Liang, H., Liu, J., Feng, X., Deng, B., Sun, T. and Shao, L. (2016). Toxicity of graphene-family nanoparticles: a general review of the origins and mechanisms. *Particle and Fibre Toxicology*, 57.

Padilla, S., Hunter, D. L., Padnos, B., Frady, S. and MacPhail, R. C. (2011). Assessing locomotor activity in larval zebrafish: Influence of extrinsic and intrinsic variables. *Neurotoxicology and Teratology*, 624-630.

Patton, E. E., Zon, L. I. and Langenau, D. M. (2021). Zebrafish disease models in drug discovery: from preclinical modelling to clinical trials. *Nature Reviews Drug Discovery*, 611-628.

Rothenbücher, T. S. P., Ledin, J., Gibbs, D., Engqvist, H., Persson, C. and Hulsart-Billström, G. (2019). Zebrafish embryo as a replacement model for initial biocompatibility studies of biomaterials and drug delivery systems. *Acta Biomaterialia*, 235-243.

Rouhi, P., Jensen, L. D., Cao, Z., Hosaka, K., Länne, T., Wahlberg, E., Steffensen, J. F. and Cao, Y. (2010). Hypoxia-induced metastasis model in embryonic zebrafish. *Nature Protocols*, 1911-1918.

Silva, N. J., Dorman, L. C., Vainchtein, I. D., Horneck, N. C. and Molofsky, A. V. (2021). In situ and transcriptomic identification of microglia in synapse-rich regions of the developing zebrafish brain. *Nature Communications*, 5916.

Simões, A. M., Venâncio, C., Alves, L., Antunes, F. E. and Lopes, I. (2021). Hydrophobic modifications of hydroxyethyl cellulose polymers: Their influence on the acute toxicity to aquatic biota. *Journal of Hazardous Materials*, 124966.

Spence, R., Gerlach, G., Lawrence, C. and Smith, C. (2008). The behaviour and ecology of the zebrafish, Danio rerio. *Biological reviews of the Cambridge Philosophical Society*, 13-34.

Stine, J. S., Harper, B. J., Conner, C. G., Velev, O. D. and Harper, S. L. (2021). In Vivo Toxicity Assessment of Chitosan-Coated Lignin Nanoparticles in Embryonic Zebrafish (*Danio rerio*). *Nanomaterials*, 111.

Talens, S. V., Arias-Alpizar, G., Makurat, D. M. M., Davis, J., Bussmann, J., Kros, A. and Kieltyka, R. E. (2020). Stab2-Mediated Clearance of Supramolecular Polymer Nanoparticles in Zebrafish Embryos. *Biomacromolecules*, 1060-1068.

Tang, G. P., Guo, H. Y., Alexis, F., Wang, X., Zeng, S., Lim, T. M., Ding, J., Yang, Y. Y. and Wang, S. (2006). Low molecular weight polyethylenimines linked by beta-cyclodextrin for gene transfer into the nervous system. *The Journal of Gene Medicine*, 736-744.

Tao, J., Wei, Z., He, Y., Yan, X., Ming-Yuen Lee, S., Wang, X., Ge, W. and Zheng, Y. (2020). Toward understanding the prolonged circulation and elimination mechanism of crosslinked polymeric micelles in zebrafish model. *Biomaterials*, 120180.

Teame, T., Zhang, Z., Ran, C., Zhang, H., Yang, Y., Ding, Q., Xie, M., Gao, C., Ye, Y., Duan, M. and Zhou, Z. (2019). The use of zebrafish (*Danio rerio*) as biomedical models. *Animal Frontiers*, 68–77.

Tomić, S. Lj., Nikodinović-Runić, J., Vukomanović, M., Babić, M. M. and Vuković, J. S. (2021). Novel Hydrogel Scaffolds Based on Alginate, Gelatin, 2-Hydroxyethyl Methacrylate, and Hydroxyapatite. *Polymers*, 932.

Tomić, S. Lj., Babić, M. M., Vuković, J. S., Djokić, L., Pavić, A., and Nikodinovic-Runić, J. (2020). Effect of composition and method of preparation of 2-hydroxyethyl methacrylate/gelatin hydrogels on biological *in vitro* (cell line) and *in vivo* (zebrafish) properties. *Journal of Polymer Research*, 305.

Truong, L., Reif, D. M., St Mary, L., Geier, M. C., Truong, H. D. and Tanguay, R. L. (2014). Multidimensional in vivo hazard assessment using zebrafish. *Toxicological Sciences*, 212-233.

Tseng, T. C., Tao, L., Hsieh, F. Y., Wei, Y., Chiu, I. M. and Hsu, S.-h. (2015). An Injectable, Self-Healing Hydrogel to Repair the Central Nervous System. *Advanced Materials*, 3518-3524.

Tzankova, V., Aluani, D., Kondeva-Burdina, M., Yordanov, Y., Odzhakov, F., Apostolov, A. and Yoncheva, K. (2017). Hepatoprotective and antioxidant activity of quercetin loaded chitosan/alginate particles *in vitro* and *in vivo* in a model of paracetamol-induced toxicity. *Biomedicine & Pharmacotherapy*, 569-579.

Usai, A., Di Franco, G., Colucci, P., Pollina, L. E., Vasile, E., Funel, N., Palmeri, M., Dente, L., Falcone, A., Morelli, L. and Raffa, V. (2020). A Model of a Zebrafish Avatar for Co-Clinical Trials. *Cancers*, 677.

Van Norman, G. A. (2019). Limitations of Animal Studies for Predicting Toxicity in Clinical Trials: Is it Time to Rethink Our Current Approach? *JACC: Basic to Translational Science*, 845-854.

Vancha, A. R., Govindaraju, S., Parsa, K. V. L., Jasti, M., González-García, M. and Ballestero, R. P. (2004). Use of polyethyleneimine polymer in cell culture as attachment factor and lipofection enhancer. *BMC Biotechnology*, 23-23.

Vimalraj, S., Yuvashree, R., Hariprabu, G., Subramanian, R., Murali, P., Veeraiyan, D. N. and Thangavelu, L. (2021). Zebrafish as a potential biomaterial testing platform for bone tissue engineering application: A special note on chitosan based bioactive materials. *International Journal of Biological Macromolecules*, 379-395.

Volz, D. C., Hipszer, R. A., Leet, J. K. and Raftery, T. D. (2015). Leveraging Embryonic Zebrafish To Prioritize ToxCast Testing. *Environmental Science & Technology Letters*, 171-176.

Vong, L. B., Kobayashi, M. and Nagasaki, Y. (2016). Evaluation of the Toxicity and Antioxidant Activity of Redox Nanoparticles in Zebrafish (*Danio rerio*) Embryos. *Molecular pharmaceutics*, 3091-3097.

Wang, Y., Zhou, J., Liu, L., Huang, C., Zhou, D. and Fu, L. (2016). Characterization and toxicology evaluation of chitosan nanoparticles on the embryonic development of zebrafish, *Danio rerio*, *Carbohydrate Polymers*, 204-210.

Wiley, D., Redfield, S. and Zon, L. (2017). Chemical screening in zebrafish for novel biological and therapeutic discovery. *Methods in cell biology*, 651-679.

Williams, D. F. (2015). Regulatory biocompatibility requirements for biomaterials used in regenerative medicine. *Journal of Materials Science: Materials in Medicine*, 89.

Wu, Q., Deng, S., Li, L., Sun, L., Yang, X., Liu, X., Liu, L., Qian, Z., Wei, Y. and Gong, C. (2013). Biodegradable polymeric micelle-encapsulated quercetin suppresses tumor growth and metastasis in both transgenic zebrafish and mouse models. *Nanoscale*, 12480-12493.

Wu, Q., Li, G., Deng, S., Ouyang, L., Li, L., Liu, L., Luo, N., Song, X., He, G., Gong, C. and Wei, Y. (2014). Enhanced antitumor activity and mechanism of biodegradable polymeric micelles-encapsulated chetomin in both transgenic zebrafish and mouse models. *Nanoscale*, 11940-11952.

Xu, C. and Zon, L. I. (2010). 9 - The zebrafish as a model for human disease. In S. F. Perry, M. Ekker, A. P. Farrell, and C. J. Brauner (Eds.), *Fish Physiology* (Vol. 29, pp. 345-365): Academic Press.

Yang, R., Xu, C., Wang, T., Wang, Y., Wang, J., Quan, D. and Deng, D. Y. B. (2017). PTMAc-PEG-PTMAc hydrogel modified by RGDC and hyaluronic acid promotes neural stem cells' survival and differentiation in vitro. *RSC Advances*, 41098-41104.

Yuan, Z., Li Y., Hu, Y., You, J., Higashisaka, K., Nagano, K, Tsutsumi, Y. and Gao, J. (2016). Chitosan nanoparticles and their Tween 80 modified counterparts disrupt the developmental profile of zebrafish embryos, *International Journal of Pharmaceutics*, 644-656.

Zang, L., Maddison, L. A. and Chen, W. (2018). Zebrafish as a Model for Obesity and Diabetes. *Frontiers in Cell and Developmental Biology*, 6(91).

Zang, L., Shimada, Y. and Nishimura, N. (2018). Development of a Novel Zebrafish Model for Type 2 Diabetes Mellitus. In *Proceedings for Annual Meeting of The Japanese Pharmacological Society WCP2018* (The 18th World Congress of Basic and Clinical Pharmacology). Japanese Pharmacological Society.

Zhang, B., Xuan, C., Ji, Y., Zhang, W. and Wang, D. (2015). Zebrafish xenotransplantation as a tool for in vivo cancer study. *Familial Cancer*, 487-493.

Zhang, W., Pan, S., Wen, Y., Luo, X. and Zhang, X. (2010). Synthesis of poly(ethylene glycol)-g-chitosan-g-poly(ethylene imine) co-polymer and in vitro study of its suitability as a gene-delivery vector. *Journal of Biomaterials Science, Polymer Edition*, 741-758.

Zhang, X., Stockhammer, O. W., de Boer, L., Vischer, N. O. E., Spaink, H. P., Grijpma, D. W. and Zaat, S. A. J. (2017). The zebrafish embryo as a model to quantify early inflammatory cell responses to biomaterials. *Journal of Biomedical Materials Research Part A*, 2522-2532.

Zhang, Y., Qin, W., Lu, X., Xu, J., Huang, H., Bai, H., Li, S. and Lin, S. (2017). Programmable base editing of zebrafish genome using a modified CRISPR-Cas9 system. *Nature Communications*, 118.

Zhao, N., Roesler, S. and Kissel, T. (2011). Synthesis of a new potential biodegradable disulfide containing poly(ethylene imine)-poly(ethylene glycol) copolymer cross-linked with click cluster for gene delivery. *International Journal of Pharmaceutics*, 197-205.

Zhou, T., Dong, Q., Shen, Y., Wu, W., Wu, H., Luo, X., Liao, X. and Wang, G. (2016). PEG-b-PCL polymeric nano-micelle inhibits vascular angiogenesis by activating p53-dependent apoptosis in zebrafish. *International Journal of Nanomedicine*, 6517-6531.

Chapter 2

The Role of Climate Change in Green Mussel Cultivation and Underlying Environmental Limitations

Chayarat Srisunont[1] and Sandhya Babel[2,*]

[1]Environmental Science and Technology Program, Faculty of Science and Technology, Bansomdejchaopraya Rajabhat University, Bangkok, Thailand
[2]School of Biochemical Engineering and Technology, Sirindhorn International Institute of Technology (SIIT), Thammasat University, Pathum Thani, Thailand

Abstract

Green mussel (*Perna viridis*) is a remarkably commercial marine bivalve species in the Indo-Pacific region. Climate change can induce variation in marine environmental parameters. Subsequently, it impacts mussel growth and survival. High air temperature can cause the rising temperature in seawater. Intense precipitation can lower seawater salinity, pH, and dissolved oxygen. Excessive nutrients and pollutants can be discharged through a high magnitude of river runoff. These robust phenomena may result in mussel deterioration. Reduction of mussel production causes loss of income in the fisheries sector, and it can have a tremendous effect worldwide. This chapter aims to critically review the effects of climate change on green mussels and the environmental factors that may influence mussel production. This study will provide information for farmers to implement appropriate cultivation methods to minimize the effect of climate change on mussel production. In addition,

[*] Corresponding Author's Email: sandhya@siit.tu.ac.th.

In: Advances in Marine Biology. Volume 6
Editors: Adam Kovács and Patrik Nagy
ISBN: 979-8-88697-996-1

other researchers can use this information to develop adaptation tools for climate change mitigation, especially in the marine ecosystem.

Keywords: marine environment, atmospheric parameters, seawater quality, climate change, aquaculture, *Perna viridis*

Introduction

Green mussel (*Perna viridis*) is a non-selected filter feeder. It can be found in estuarine and coastal line areas. The mussels are farmed extensively throughout the world. Most commercial farms are in Thailand, the Philippines, India, Indonesia, and New Zealand (FAO, 2021). According to a report in Fisheries Statistic of Thailand 2019 (Information and Communication Technology Center, 2021), green mussel production was 38,000 tons which value 524.10 million Baht (15.72 million USD). It can be accounted for the largest amount of shellfish production in Thailand.

Climatic change, caused by global warming and ozone depletion, can induce variation in marine environmental factors. Subsequently, green mussel mortality may occur. Anthropogenic activities release tremendous carbon dioxide in the atmosphere, causing the rising global temperature, ocean acidification, and climate change, strongly affecting the marine ecosystem. The Intergovernmental Panel on Climate Change: IPCC (2021) forecasted that surface seawater temperature would be increased 1.0-5.7 °C and the pH of seawater would decrease by 0.1-0.4 units at the end of the century. Recently, it was found that pH was already below 8.06 (European Environment Agency, 2021). The frequency and magnitude of large storms and turbulent weather have increased. Extreme precipitation events can rapidly reduce salinity and enhance nutrient loading in estuarine and coastal areas (Doney et al., 2012). These robust phenomena can result in the degradation of marine living organisms and the marine culture of green mussel farms, one of the main commercial food industries in the Indo-Pacific region. Finally, all countries in this region will lose much income in the fisheries sector due to climate change.

Although the green mussel has been a remarkable species due to its effectiveness in filtration and wide range of environmental resistance (Rajagopal et al., 2006), it still has a threshold limit. Thermal stress can cause mussel mortality (Goh and Lai, 2014; Wendling et al., 2013). Lowered pH in seawater caused by the dissolution of carbon dioxide may affect physiological and immune responses in green mussels (Sivalingam, 1977). It was also found that lowered pH causes growth reduction and increased respiration rate in the Chilean mussel *Mytilus chilensis* (Duarte et al., 2014) and the Mediterranean mussel *Mytilus galloprovincialis* (Gestoso et al., 2016). Degradation of mussel hemocytes also increased with decreased salinity (Wang et al., 2012). These environmental limitations on green mussels have been researched; however, none discuss the mussels' growth and survival under environmental conditions fluctuation caused by climate change. Hence, this chapter aimed to critically review the effects of climate change on green mussels, environmental factors influencing mussel production, and appropriate mussel cultivation method during climate change events. Knowledge forms these issues can assist scientists in developing climate change mitigation tools for mussel farms and other fisheries sectors. It can also help the farmers forecast mussel production and prepare their adaptation strategies during turbulent weather caused by climate change.

Biology of Green Mussels

Green mussels (*Perna viridis*, Linnaeus 1758) are marine species in the phylum Mollusca, class Bivalvia, order, Mytiloida, family Mytilidae, and have synonym names: *Mytilus viridis* and *M. smaragdinus* (Tuaycharoen et al., 1998; Vakily, 1989).

Morphology

Green mussel is a large mussel, 8-10 cm in length, occasionally reaching 16.5 cm. The shell tapers to a sharp, downturned beak and has a smooth surface covered with a periostracum (skin) that can be vivid green to dark brownish-green near the outer edge and olive-green near the attachment point. The ventral margin of the shell is straight or weakly concave. The interior of the shell valves is shiny and pale, bluish-green. The ridge which supports the ligament connecting the two shell valves is finely pitted. The beak has interlocking teeth: one in the right valve and two in the left. The wavy posterior end of the pallial line and the large kidney-shaped adductor muscle are diagnostic features of this species (Figure 1).

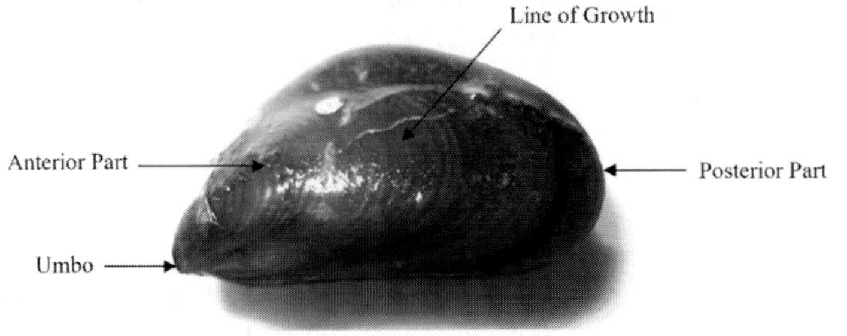

Figure 1. Morphology of green mussel's shell.

Reproduction

Green mussels are separate and have external fertilization. Spawning generally occurs twice a year, between early spring and late autumn. However, in the Philippines and Thailand, spawning takes place year-round. Fertilized eggs develop into larvae and remain in the water column for two weeks before settling as juveniles. When the larvae metamorphose becomes small mussels (spat), they grow byssus threads, which attach to a suitable firm object. They can move slowly through attaching and detaching byssus threads to attain a better life position. Sexual maturity typically occurs at 1.5-3.0 cm shell length (2-3 months of age). The life span of green mussel is typically 2-3 years (Vakily, 1989).

Growth Rate

The growth rate is influenced by environmental factors such as temperature, salinity, and food availability. The first-year growth rate varies between locations and ranges from 4.97 cm/year in Hong Kong (Lee, 1985) to 12 cm/year in India (Rajagopal et al., 1998b). Growth of *Perna viridis* or green mussel is most measured as an increase in length. "Length" in this context is the maximum distance along the umbo to the posterior axis of the animal (Figure 1). In Thailand, the growth rate is around 1 cm/month in shell length (Teeramaethee, 1998). The marketable size is 5-6 mm shell length or after 5-6 months culture.

Distribution and Habitat

Green mussels are the tropical native coastal in the Indo-Pacific region. Their habitat is in the intertidal zone with high salinity (25-30 psu or practical salinity unit). However, they prefer always to submerge (Dore, 1991). Green mussels can form dense populations (up to 35,000 individuals/m^2) on various structures, including vessels, wharves, mariculture equipment, buoys, and other hard substrata in the estuary or coastal area (Vakily, 1989). The green mussel can aggregate a large cluster as gregarious marine life by well-developed byssal apparatus. It is susceptible to overgrowth from other fouling organisms, but it is not easy to detect despite its vivid green appearance.

Green Mussel Predators

Green mussel is another commercial bivalve that feeds on phytoplankton. Many shellfish predators need to be considered by the mussel growers, such as crustaceans, gastropods, echinoderms, fishes, and sea birds. Many large crustaceans, including crabs and lobsters, prey on shellfish as a major food source. Especially, crabs are major predators for all shellfish species, including green mussels. They can open the shell by crushing it or prying it open with their claws. Not only the crustacean and gastropods like moon snails and whelks are significant predators of cultured bivalves, but also they can drill through shells using their radula, a rasping appendage similar to a rough tongue. Octopus and squids also can pray bivalves by opening bivalve shells with their powerful tentacles, exposing the shellfish to their sharp beaks. Echinoderms, which prey on shellfish, are mostly starfishes. They slowly open the shell with their arms and insert their stomach through the narrow gap, digesting the shellfish inside its shell. Then they suck the shellfish meat into their body. Many fish such as parrot fish, small sharks, stingrays, skates, and groupers also can crack open the shells of shellfish in their jaws. Sea birds, such as waterfowl, herons, seagulls, and cranes, can crack open small shellfish with their beaks (NRAC, 1993). These marine animals should be considered as green mussel consumers which can reduce the number of commercial mussels.

Feeding and Filtration Rate

Green mussels are attached to a hard substrate in intertidal zones. However, they prefer to be submerged. Growth rates tend to be higher for fully submerged animals. They are an efficient filter feeder, feeding on small zooplankton, phytoplankton, and other fine organic material (Sivalingam, 1977). By their filtration activity, all suspended solids will go through green mussels' mantle cavity and gill. In a polyculture with the fish farm, Gao et al. (2006) found that green mussels can remove all organic waste such as particulate organic matters (POM), fish feed, and fish feces through their filter-feeding. However, the terrestrial suspended solid can be rejected and released as pseudofaeces without digestion, and the integrity of green mussel increase with higher phytoplankton density or chlorophyll-*a* concentration (Rajagopal et al., 2006). In addition, green mussel filtration rates were associated with the mussel size and density. Research on the mussel filtration model by Tantanasarit et al., (2013) demonstrated that mussel filtration rates were more effective in small mussels. Low mussel densities provide large volumes of seawater/mussel; subsequently, high filtration rate. The maximum filtration rate can be reached at 10.37 L/hr/g DW tissue, and the first-order kinetics model can determine it.

Likewise, other researchers showed that blue mussel (*Mytilus edulis*) effectively filters all particulate matter in seawater. However, the real food source is phytoplankton. The investigation in integrated cultivation mussel with western king prawn in a close system by van Khoi and Fotedar (2012) demonstrated that the mussels could grow and survive in the close integration system and effectively remove total suspended solids, total nitrogen, and total bacteria from the cultured environment. Liutkus et al., (2012) also found that the blue mussel can feed all particles in seawater, including wastes from shrimp and fish farms. However, the mussel integrity was reduced because its growth was reliable in the local phytoplankton community. During algal blooms, there were phytoplankton cells in the mussel filtration and digestion process. It showed that although the mussels can consume all suspended particles, only phytoplankton will be digested and enhance mussel growth.

Another research by Pernet et al., (2012) also found that the food source of mussels is all terrestrial organic matter and bacteria during non-bloom phytoplankton periods. However, the high growth rate in bivalves is observed when water mainly contains phytoplankton, especially diatoms. POM and chlorophyll-*a* in the bivalve farming area are lower than those recorded outside the farming area (Pernet et al., 2012). The results also indicated the contribution of diatoms to the diet of mussels. Moreover, mussels also can remove bacteria from seawater. However, bacteria likely represent a minor food source for bivalves. Thus, it can be concluded that mussels feed more on diatoms (or marine phytoplankton) and less on terrestrial organic matter and bacteria.

Effect of Climate Change on Green Mussel

It has been known that the dramatic increase of carbon dioxide in the atmosphere caused by anthropogenic disturbance, especially deforestation and combustion of fossil fuels, results in global warming and climate change (IPCC, 2021). Weather changes can cause environmental fluctuation in estuarine and coastal areas, such as a rise in seawater temperature, ocean acidification, and decreased salinity. Rising air temperature due to a tremendous amount of greenhouse gas (GHGs) can increase surface temperature. High volumes of carbon dioxide in the atmosphere dissolved in seawater can decrease pH due to the formation of carbonic acid. Weather disturbances such as storms, heavy rainfall, floods, and river runoff cause rapid salinity reduction, lower dissolved oxygen (DO), turbidity, and high loading of nutrients, especially in estuarine and coastal areas. These environmental stresses affect marine living organisms, including green mussels.

Air Temperature Rising

Climate change causes air temperature variations leading to the fluctuation in seawater temperature. By the end of the century, the year 2100, global surface temperatures are predicted to increase approximately 1.2-5.7°C, increasing on average 0.1-0.2°C per decade (IPCC, 2021). An increase in air temperature causes an increase in surface water temperature due to the enhanced heat content of water by solar radiation (Morrill et al., 2001; Harvey et al., 2013). This phenomenon can be described by linear regression (Johnson 1971; Song et al., 1973; Harvey et al., 2013; Srisunont et al., 2022). Unstable air temperature can cause thermal stress affecting physiological and biochemical changes in green mussels.

Global climate change can cause warm and cold thermal stress on green mussels. Long-term climate warming can trigger mussel growth and survival. At a high temperature, mussels' immune response as esterase, reactive oxygen species, lysosome content, and phagocytosis can be reduced (Wang et al., 2011). Deformation of gonad can be found (Sreedevi et al., 2014). Biochemical components, physiological response, hemocytes production, and clearance rate can be declined (Goh and Lai, 2014). Moreover, the experiment on climate change–related stressors in green mussels by Wendling et al., (2013) indicated that rising seawater temperature caused mussel mortality. The mussels survived only 25% to 75% when the seawater temperature reached 34°C, and they were completely die-off at 38°C. On the contrary, cold air temperature during the winter caused mussel mortality in some areas. Research by Firth et al., (2011) demonstrated that mass mussel mortalities were observed in Florida, the USA, when the air temperature was below 2°C for 6 hours during low tide. Low air temperature causes a reduction of seawater temperature; subsequently, low phagocytosis and increased oxidative metabolism (Donaghy and Volety, 2011). Therefore, it can be summarized that thermal stress caused by climate change leads to mussel mortality due to the decline of immune function and hemocytes production.

Ocean Acidification

Oceans are important as a carbon sink. When carbon dioxide dissolves in seawater, it changes to carbonic acid, decreasing both pH and availability of carbonate ions. It is predicted that at the end of the century (2100), the pH level would be 8.05-7.65 or decrease by 0.1-0.4 units (IPCC, 2021). Accordingly, this condition of ocean acidification can induce physiological and biochemical stress in marine organisms, particularly green mussels. At present, information on the effect of ocean acidification on green mussels' growth and survival is limited. Sasikumar and Krishnakumar (2011) found that the mussel integrity was reduced when the pH level in seawater at the cultivation area was lowered to 7.4. However, some research has been done on other mussel species. For example, research on blue mussel (*Mytilus edulis*) by Gu et al., (2019) revealed that lowered pH level from 8.1 to 7.3 caused an increased excretion rate and reduced clearance rate, absorption efficiency, respiration rate, and scope for growth of the mussel. Sui et al., (2016b) found that after the mussels, *Mytilus coruscus,* were exposed to seawater with pH 7.3, their immune responses such as hemocytes, phagocytosis, esterase, and lysosomal content were decreased. Nonetheless, the effect on their clearance rate, absorption efficiency, respiration rate, ammonia excretion rate, and scope for growth (SFG) was not found (Sui et al., 2016a). Similarly, Wu et al., (2016) reported that at pH 7.3 and exposure for 7 days, the hemocytes mortality increased in the mussel *Mytilus coruscus*. When the experiment duration was extended, the effect of pH was significantly clearer. Gestoso et al., (2016) found that in the Mediterranean mussel *Mytilus galloprovincialis*, reduced growth rate and increased respiration rate were observed when exposed to seawater pH 7.65 for 22 days.

Moreover, the higher dissolution of carbon dioxide (CO_2) in seawater can cause a reduction in the carbonate ion concentration of seawater. Reduction in calcification ability in juvenile blue mussel (*Mytilus edulis*) (Fitzer et al., 2014) and the Chilean mussel *Mytilus chilensis* (Duarte et al., 2014) were found when the mussels were introduced to high CO_2 dissolution (1,000 ppm) for 6 and 2 months, respectively. In addition, ocean acidification caused by the high dissolution of CO_2 directly affects the mussel and their predator. Research on predator-prey interaction between the muricid gastropod *Thais clavigera* and the mussel *Brachidontes variabilis* by Xu et al., (2017) demonstrated that the gastropod performed longer pray searching time at high CO_2 dissolution (1,250 ppm). It may be due to weakening a chemosensory response to prey cues, resulting in the inability to distinguish between empty shells and live individuals.

Accordingly, the research studies imply that during ocean acidification caused by climate change, green mussels are impacted by decreased immune function, physiological response, and calcification ability. Nevertheless, research on green mussels is still required because it may be a species-specific response.

Extreme Weather

Weather disturbances such as storms and heavy rainfalls cause rapid salinity reduction, especially in estuarine and coastal areas (Doney et al., 2012). It can strongly affect green mussel growth and survival. Reduction of byssus threads and respiratory rate can be found when salinity is lowered to 15 psu (Huhn et al., 2016). The mussels tend to close their valves and die when exposed for 120 hours (McFarland et al., 2013). This stress resulted from the degradation of immune functions. Mussels reduce the production of haemocytes, phagocytosis, esterase, reactive oxygen species, and lysosomal content when salinity becomes lowered (Wang et al., 2012). They could not recover immunity function in a short time. The oxygen consumption is reduced, causing insufficient oxygen to survive, and finally, leading to mussel mortality (Wendling et al., 2013).

Moreover, the mussel's condition index and calcification rate were lowered when salinity decreased. Bivalves reduced calcification rate through physiological constraints associated with osmotic stress (Sanders et al., 2018). Consequently, the growth of mussels' declined.

Heavy rainfall and storms cause floods and tremendous river runoff leading to fluctuation in dissolved oxygen (DO), water turbulence, and turbidity. This phenomenon strongly affects green mussels' physiological functions. In the aquatic system, hypoxia, or reduction of DO that reaches the threshold limit of living organisms, can be found after intense rainfall due to decomposition of organic matters and pollutants flushed from the river (Eglash, 2005). This phenomenon causes a decline in immune function and hemocytes production in green mussels (Wang et al., 2012). Water turbulence and turbidity influences the ability of mussel filtration. Mussels feed on seston, which includes phytoplankton and other suspended organic matters. However, they reject some inorganic particles without passing through the digestive tract as pseudofaeces (Wong and Cheung, 1999). Under high turbidity, their gill filament can be damaged (Shin et al., 2002). Consequently, it affects feeding mechanisms, and eventually, the filtration rate decline.

Global warming and climate change cause more frequent and extreme precipitations, bringing more fresh water and nutrients loading to estuarine and coastal areas. This high loading of nutrients enhances phytoplankton abundance, which is green mussels' main food source. For instance, climatic conditions such as El Nino and La Nina reflect phytoplankton and organic matter variation. Alfaro et al., (2010) found that surface seawater temperature was higher during the El Nino period with less wind flow. It causes changes in wind-driven Ekman transport; subsequently, reducing upwelling, which brings nutrients from the seafloor associated with phytoplankton growth. They also found low green mussel spat during this period due to low available food. In contrast, there were more frequent storms and strong wind during the La Nina period. This phenomenon caused upwelling, which distributed nutrients from the seafloor to surface water. Then, it enhanced phytoplankton growth, a limiting factor of green mussel growth and survival. In opposite, the high value of nutrient concentration can cause physical change and mortality of green mussels. The study on the effects of ammonia/ammonium on tolerance and byssogenesis in green mussels by Reddy and Menon (1979) demonstrated that young mussels (shell length 20-24 mm) could detach from the substrate when exposed to ammonia concentration of 2.26 mg/L for 96 hours. Moreover, the mussels started to close their valve after being exposed to the ammonia concentration of 11 and 5.9 mg-N/L for 48 and 96 hours, respectively. Another research on the influence of ammonia concentration on green mussels by Fang et al. (2008) also found that 2.0 mg-N/L of ammonia concentration significantly reduced the lysosomal integrity of green mussels. It can be concluded from cited literature that although enhancement of nutrients through river runoff during extreme weather can provide food abundance for green mussels, it can be toxic to the mussel if the concentration is over the threshold limit.

Environmental Factors and Impact on Green Mussels

Although green mussels are remarkable species in their ability to withstand environmental fluctuations (Rajagopal et al., 2006), some limitations remain. Green mussels attach themselves with the hard substrate by byssus threads, submerge in the intertidal zone, and feed on plankton and other microscopic sea creatures, free-floating in seawater. The biochemical component in green mussels changed with seasonal variations of environmental conditions (Verlecar et al., 2008). Marine environmental factors such as temperature, pH, salinity, dissolved oxygen (DO), turbidity (total suspended solids: TSS), and food availability (chlorophyll-*a* or phytoplankton concentration) (Table 1) influence the growth and survival of green mussels.

Temperature

Seawater temperature influences the growth and survival of green mussels. Normally, green mussels are found in estuarine and coastal areas with seawater temperature in the range of 11-36 °C (Firth et al., 2011; Gao et al., 2008; Rajagopal et al., 1998b; Sivalingam, 1977; Trottet et al., 2007; Wang et al., 2011; Wendling et al., 2013; Wong and Cheung, 2001a). They can resist a wide range of surface temperatures; however, they still have limitations (Table 1).

Table 1. Green mussel physiological response to variation of seawater factors

Seawater Parameters	Green Mussel Physiological Response	Reference
Temperature (°C)		
38.00	100% die-off	Goh and Lai (2014)
36.00	Survive	Wong and Cheung (2003)
34.00	25-75 die-off	Wendling et al., (2013)
33.00	100% larvae mortality	Nair R and Appukuttan (2003)
31.00	Reduce respiration	Nair R and Appukuttan (2003)
25.00	Induce spawning	Wong and Cheung (2003)
14.00	Die-off found	Firth et al., (2011)
11.00	Lowered condition index	Hickman et al., (1991)
10.00	50% die-off	Sivalingam (1977)
8.00	100% die-off	Goh and Lai (2014)
pH		
9.00	50% die-off	Sivalingam (1977)
8.65	Survive	Sasikumar and Krishnakumar (2011)
7.54	Survive	Sasikumar and Krishnakumar (2011)
7.40	Lowered mussel integrity	Sasikumar and Krishnakumar (2011)
3.50	50% die-off	Sivalingam (1977)
Salinity (psu)		
80.00	50% die-off	Sivalingam (1977)
65.00	Survive	Sivalingam (1977)
25.00	Survive	Gao et al., (2008)
24.00	50% die-off	Sivalingam (1977)
15.00	100 die-off	Roungjai (1999)
DO (mg/L)		
8.00	Survive	Wong and Cheung (2001a)
4.00	Survive	Wu and Lam (1997)
3.00	Fewer byssus thread	Wang et al., (2009)
2.50	Low lysosomal integrity	Fang et al.,(2008)
1.50	Reduce immunity	Wang et al., (2012)
0.50	75% die-off	Wendling et al., (2013)
TSS (mg/L)		
1,200	100% die-off	Shin et al., (2002)
500	Cilia depletion	Cheung and Shin (2005)
250	Survive	Cheung and Shin (2005)
Chlorophyll *a* (μg/L)		
80	Pseudofeaces production	Rajagopal et al., (2001)
18	Survive	Rajagopal et al., (1998a)
1	Survive	Rajagopal et al., (1998a)

When seawater temperatures were below 14°C, the mussels die-off was found (Firth et al., 2011). At 11°C, the reduction of the condition index of green mussel was reported (Hickman et al., 1991). At 10°C caused 50% of green mussels die-off in 2 weeks (Sivalingam, 1977). Finally, a complete die-off was found at seawater temperature at 8 °C (Goh and Lai, 2014).

The growth rate started to increase with the increased seawater temperature, and the temperature at 25°C induced spawning of the green mussel. Other studies also stated that a high growth rate is related to high water temperature due to enhancement of feeding and absorption ability (Alfaro et al., 2001; Hickman, 1979; Rajagopal et al., 1998b; Wong and Cheung, 2003). However, thermal stress can cause mussel mortality. At seawater temperature reaching 31°C, green mussels reduced respiration. Then, green mussel larvae mortality was found when seawater temperature was higher than 33°C (Nair R and Appukuttan, 2003). For adult mussels, only 25-75% of mussels can survive when seawater temperature reaches 34°C (Wendling et al., 2013), and none of the mussels can survive at 38°C (Goh and Lai, 2014). These mortalities were caused by lysosome reduction, degradation of mussels' integrity (Wang et al., 2011), growth rate reductions, and increased respiratory rate (Gestoso et al., 2016) when seawater temperature was beyond the threshold limit.

pH

The intertidal zone is the natural habitat of green mussels; pH value is varied and affects mussels' physiological responses (Cebu and Orale, 2017). The increased pH level results in high mussel integrity. The suitable pH level for mussel growth is 7.54-8.65 (Sasikumar and Krishnakumar, 2011). Extreme high or low pH can affect the balance of sodium and chloride in the blood of mussels; subsequently, respiratory failure, loss of osmotic regulation, and death. When the pH level is higher than 9, it is harmful to green mussels. This condition causes a 50% die-off (Sivalingam, 1977). When pH is lower than 7.4, it causes a reduction of mussels' condition index. This condition may support green mussels but not recruitment due to low growth rate and integrity (Soon and Ransangan, 2016). When the pH level is lower than 6.9, this seawater property cannot support green mussel growth and survival. Finally, only 50% of the mussels can survive when the pH level drops to 3.5 (Table 1) (Sivalingam, 1977).

Salinity

Green mussels are marine animals. They generally grow with salinity in the range of 25-35 psu (Hickman et al., 1991; Gao et al., 2008, Rajagopal et al., 1998b; Trottet et al., 2007; Wong and Cheung, 2001a). Sivalingam (1977) stated that the optimum salinity for normal growth was between 27 and 65 psu. When salinity was lower than 24 psu or higher than 80 psu, green mussels could stand for and die-off 50% after 2 weeks. Green mussels exhibit their metabolism at low salinity by closing the valve to stay alive, but not for a long period. Then, green mussels become mortal when salinity is lower than 15 psu (Roungjai, 1999) because they reduce respiratory rate (Wendling et al., 2013) and immune functions (Wang et al., 2012). In short, green mussels prefer to live at salinity higher than 25 psu; however, they can die at salinity over 80 psu.

Dissolved Oxygen (DO)

Dissolved oxygen (DO) is an influencing environmental factor affecting green mussel growth. Green mussels can be found in seawater that contains DO 4-8 mg/L (Rajagopal et al., 1998b; Sivalingam, 1977; Wong and Cheung, 2001a). When DO concentration is lower than 4 mg/L, green mussels increase their respiration enzyme due to biochemical adjustment (Wu and Lam, 1997). Mussels also reduce lysosomal integrity when DO decreases to 2.5 mg/L (Fang et al., 2008). Another research about the physical effect on green mussels at low DO was observed by Wang et al., (2009). They demonstrated that green mussels produced very few byssus threads which also were very short and thin at reduced oxygen levels to 1.5-3 mg/L. This research showed that the minimum amount of oxygen that byssus production required for settlement on a substrate is 3 mg/L. At extreme low DO of 1.5 mg/L, green mussels will reduce immune functions and increase hemocytes mortality (Wang et al., 2012). Finally, mass mortality can be found when DO concentration drops to 0.5 mg/L (Wendling et al., 2013).

Turbidity

Even though green mussels can filter all suspended particles in seawater, there is a limitation. In natural conditions, green mussels can generally grow with total suspended solids (TSS) in the range of 10-15 mg/L (Gao et al., 2008; Wong and Cheung, 2001a). In turbid water with high TSS concentration, all suspended particles mostly come from soil erosion or are resuspended from sediment on the seafloor. These particles are less nutritious than phytoplankton, and green mussels do not want them. Green mussels can pre-ingestive reject these particles by increasing pseudofaeces production (Wong and Cheung, 1999). The proportion of filtered matter rejected as pseudofaeces before ingestion remained constant, averaging 0.87 at TSS concentrations above 40 mg/L (Tamtin et al., 2004).

Moreover, suspended particle size and concentration can cause gill damage in green mussels. Research by Cheung and Shin (2005) found no effect on cilia depletion when TSS concentration was lower than 250 mg/L. When TSS reached 500 mg/L, there was 50% cilia depletion within 7 days and reached 90% depletion at day 14. Complete cilia depletion was achieved at TSS concentration 1,000 mg/L. Shin et al. (2002) also found that green mussels could tolerate a high level of TSS up to 1,200 mg/L; however, in the long term, high TSS could damage mussels' morphology such as gill filaments and reduce mussels' growth rate and respiration.

Consequently, a decline in the filtration rate occurred. Results suggested that very high TSS caused physical overloading of feeding mechanisms in green mussels (Hawkins et al., 1999). Thus, the mussels prefer clear seawater that contain enough food source. They can hardly be cultured in turbid water that contains many suspended particles. In this condition, they can survive; however, their gills will be damaged long-term. Consequently, the filtration performances are reduced, and finally, mussels' mortality can be found.

Food Available

Food availability is another considerable parameter influencing the feeding behavior of green mussels. Green mussels are suspended filter feeders, so their food sources are all seston as phytoplankton, zooplankton, suspended particles, and organic matter. However, green mussels are more reliable on phytoplankton than other suspended particles due to higher nutrition. Rajagopal et al., (1998b) stated that mussel larvae availability in the coastal water depended more on the phytoplankton community. In natural conditions, green mussels can be found in seawater with phytoplankton 1,250-22,500 cells/mL or chlorophyll-*a* concentration at 1-18 μg/L (Rajagopal et al., 1998a; Trottet et al., 2007). Mussels can achieve high filtration and ingestion rate with increased algal concentrations. However, pseudofeaces production started when algal concentration reached 10^5 cells/mL or chlorophyll-*a* at 80 μg/L (Rajesh et al., 2001). Organic matter, including zooplankton, is also another food source of green mussels. Many researchers indicated that green mussels could filter organic matter from the water column, and the ingestion rates increased with the increase of organic matter (Hawkins et al., 1998; Wong and Cheung, 1999; Wong and Cheung 2001b). Thus, the amount of phytoplankton and suspended organic matter in the seawater column can limit green mussel growth and survival due to food availability and filtration ability.

Green Mussel Cultivation under Climate Change Events

Under climate change events, rising air temperature, heavy rainfall, frequent storms, and ocean acidification can affect green mussel culture. This section discusses the effect of climate change on cultivation methods and how to mitigate the climate crisis. Green mussels can form dense populations (up to 35,000 individuals/m^2) on various structures, including vessels, wharves, mariculture equipment, buoys, and other hard substrata in the estuary or coastal areas (Vakily, 1989). The green mussel can aggregate a large cluster as gregarious marine life by well-developed byssal apparatus. It is susceptible to overgrowth from other fouling organisms, and it is not easy to detect despite its vivid green appearance. Hence, to culture mass mussels, farmers should be concerned about cultivature site, farm method, seeding, mussel growth, disease, and poisoning.

Site Selection

Site selection is the most important farming technique to produce green mussels. The suitable areas for farming are open sea, coastal, and estuarine areas that are free from strong wave action. A moderate water current is required to bring planktonic food and carry feces, pseudofaeces, and silt in the culture area. The suitable water current should be in the range of 0.17-0.25 m/s at flood tide and 0.25- 0.35 m/s at ebb tide. Seawater quality ideal for mussel culture must be low turbidity, an abundance of plankton (or chlorophyll-*a* concentration in the range of 17-40 μg/L), and salinity in 27-35 psu, and temperature in the range of 26-32°C. Cultivation sites should be far from domestic, industrial, and sewage pollution (Alloycious, 2016).

Green mussel farming in the open sea and coastal areas should be located above 5 m depth without strong wave action, less turbulent, and high phytoplankton productivity. The suitable cultivation methods for these areas are the raft method and longline culture. Juvenile mussels naturally settle on cultivation ropes. Green mussels effectively grow and reach market size within 6-7 months due to tremendous food consumption through a large volume of seawater. However, disadvantages of these farm areas are the poaching and unpredicted climate changes.

Mussel farms are also effectively operated in estuarine areas that are less turbulent and shallow (<4 m). Estuarine areas are more preferred in primary production than open sea and coastal areas. Consequently, high mussel production can be found. Mostly, mussels reproduction and mussel seeds are found in estuarine areas. Then, these small mussels will be transported to open sea or coastal areas and will be cultured until they reach market size. The suitable cultivation methods in estuarine areas are the rack method and horizontal culture. However, salinity fluctuation during monsoon season and pollution discharge from domestic and industrial wastewater can lead to low mussel production in estuarine farming.

During climate change events, each cultivation area is affected by different parameters. The most influential factors affecting mussel production in the open sea are nutrients and pH; in estuaries are temperature and salinity (Alosairi et al., 2019; Srisunont et al., 2022). These seawater parameters can be fluctuated by climate change events. A study on climate influence on green mussels by Srisunont et al., (2022) revealed that high air temperature and intense rainfall caused an increase in seawater temperature and a decrease in salinity, respectively; subsequently, low mussel production. Their model simulations also demonstrated that intense rainfall significantly affected mussels cultured close to the river mouth. It demonstrates that locations with less riverine influence can provide higher mussel productivity. Estuaries are more sensitive to climate change than the open sea. Therefore, cultivating in open sea areas or a large distance from the river mouth can effectively produce green mussels under extreme weather events caused by climate change.

Farming Methods

There are 5 mussel farming methods: rack method, raft method, longline method, horizontal culture, and bouchot culture (Alloycious, 2016). Each method is suitable for different cultivation sites. The climate change effect and details of each method are as follows.

Rack Method

The rack method is suitable for estuaries and shallow waters. It is an adaptation of the hanging culture method. The construction is wooden rack poles (bamboo or casurina) and driven into the bottom vertically two to four meters apart. Then, these poles are connected by horizontal poles, which contain tying and lashing nylon/coir ropes. These ropes support the suspended strings of clutch (chaplets), which are placed about 30 cm apart (Figure 2). The horizontal poles should be above the water level at high tide, and seeded ropes are suspended from the same (Alloycious, 2016). Racks are used extensively for spat collection and the growth of young mussels.

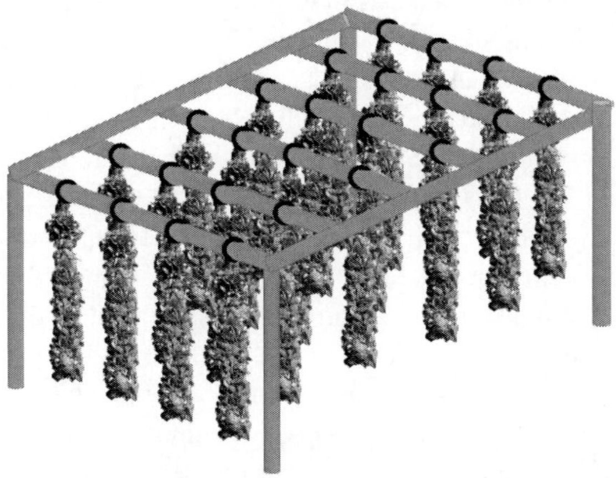

Figure 2. Rack method.

The rack structure and green mussels' integrity can deteriorate during climate change events. Strong wind, waves, and intense rainfalls influence the structural strength of the rack method. Water turbulence in estuarine areas brings freshwater, organic matters, and pollutants through river runoff. This phenomenon affects green mussel growth and survival due to salinity and dissolved oxygen depletion. Moreover, high air temperature causes a rise in seawater temperature; subsequently, thermal stress in green mussels cultured in shallow water (Srisunont et al., 2022). Several ways can minimize these climate change effects. The horizontal and vertical poles can be composite with concrete instead of woods to strengthen the structure. The heavy anchors should be added to each side of the rack. The cultivation period should avoid the monsoon season. A deeper site should be selected to prevent thermal stress during low tide. These adaptations can assist farmers who are willing to use this cultivation method.

Raft Method

Raft culture is one of the commercially important methods of intensive aquaculture. This method is ideal for the open sea and coastal areas. Square or rectangular rafts are fabricated with sturdy bamboo or casuarina poles. Buoyancy for the raft is provided by tying 5 barrels of 200 L capacity, one each at the four corners and one in the middle (metal oil barrel painted with anticorrosive paint or synthetic material). The ideal size of the raft is 5x5 m. The rafts are positioned at a suitable site in the sea using 50-100 kg of iron, granite, or concrete anchors. Three seeded ropes can be suspended from the one-meter area of the raft (Alloycious, 2016).

Raft culture is similar in many respects to longline culture, except that the raft commonly swings on a single mooring. As with longline culture, ropes are suspended from the raft. This technique has far less scope for mechanization; most harvesting is done by hand, and the operation is generally smaller in scale than longline.

Figure 3. Raft method.

A new raft culture method using recycled materials for green mussel production was developed by the Sriracha Fisheries Research Station (SFRS), Thailand (Figure 3). This method is convenient and comfortable to harvest system. It can resist wind and wave action and provide some of the benefits of artificial reefs. The raft culture system is predominantly made from polypropylene and polyethylene components, mostly obtained second-hand from industry and old trawl nets. These farming techniques make 1,600 m^2 of mussel raft, 50 m lengths of 16 mm polypropylene rope are tied together to create a grid 40 mm length and width with ropes spaced at 1 m intervals. The outside perimeter of the raft is tied with 20 mm polypropylene rope. Sealed plastic containers (20-30 L) are tightly tied to each knot in the grid with 6-8 mm polypropylene rope to provide flotation for the raft (Tunkijjanukij and Intarachart, 2007).

This cultivation method is suitable for open sea or coastal areas with less climate change impact than estuary areas (Srisunont et al., 2022). This cultivation method was mostly employed in the area far from the river mouth. Therefore, there is less salinity fluctuation and contamination of pollutant from river discharge during heavy rainfalls. These cause suitable seawater qualities for mussel growth and survival even when heavy storms occur. However, the raft structure may be destroyed if frequent tropical storms. Strong winds and waves can decrease the strength of the anchors and move the raft away. Hence, to mitigate this effect, more anchors should be added, and the raft should be placed in a location where there is steady wind.

Longline Method

The longline culture is a modification of raft culture. This method is considered ideal in unprotected open sea conditions (offshore culture). The basic feature of a longline unit is a series of styrofoam floats arranged in a row. The main line is a synthetic rope of 16-20 mm diameter. The longline is supported by 200 L barrels tied to it and spaced at 5 m. The long lines and barrels are anchored both ends using concrete blocks and nylon ropes. Seeded ropes are suspended in the longline (Alloycious, 2016). These support many culture droppers (Figure 4). The number of droppers used varies with hydrological and biological conditions but is normally 400-500 per longline. The longline can hang chaplets, pocket panels, lantern baskets, or spat collectors.

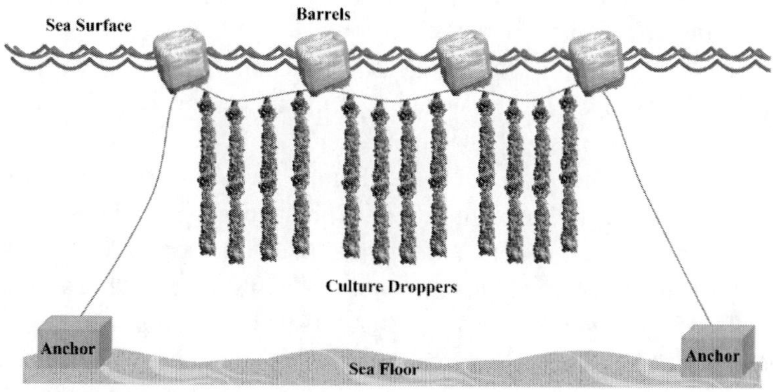

Figure 4. Longline method.

This practice is the most recent development for mussel culture. Although various types exist, subsurface longlines have been developed in France to resist storm and wave effects and are particularly adapted to areas showing high tidal cycles. This technique allows highly mechanized culture and yields 18-20 tons/ha/yr. A multi-longline system has also been developed in Norway and Sweden, using 7-9 headlines. Temporarily submerging mussel longlines is part of a farming practice designed to lower the crop below surface ice during winter months (Canada). Control of buoyancy is necessary for this system. Floats are connected by horizontal lines that support many vertical ropes where mussels are grown. Seed is either caught on collector ropes hung from the floating lines or gathered from the natural settlement in intertidal areas. Thinning and reseeding onto grow-out ropes or into stockings are carried out until the mussels reach marketable size (Barnabe, 1994; FAO, 2009).

Climate change may not affect the longline method. The method is employed in open sea areas with low climate change impact. The influencing parameters in these areas are nutrients and pH (Srisunont et al., 2022), which were forecasted to be in an optimum range for mussel growth and survival (IPCC, 2021). In addition, the structure is strong enough to resist strong wind and waves during weather turbulence. Hence, it can be assumed that the longline method is the best option to culture green mussels during climate change.

Horizontal Culture

This method is ideal in shallow areas with a minimum level of the water column. Seeded ropes are suspended by tying upward by ropes to horizontal poles, but both the ends will be stretched and tied in vertical poles erected on opposite sides in the farm structure (Figure 5). Most farmers follow this method in estuaries (Alloycious, 2016). Climate change can affect structural strength, seawater parameters, and thermal stress like the rack method. This method is usually employed in the pond or shallow water with less than 2 m depth. This area is very sensitive to thermal stress caused by climate change (Srisunont et al., 2022). Therefore, this method may not suit green mussel cultivation because mass mortality may occur during climate change events.

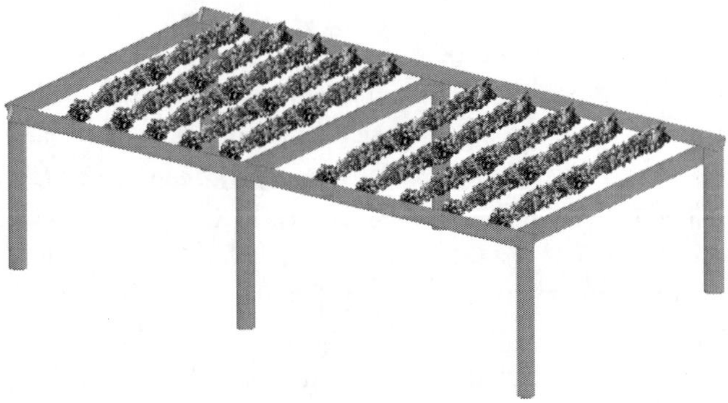

Figure 5. Horizontal culture.

Bouchot Culture

"Bouchot" culture is mainly undertaken in France. It is also called the "pole culture" or stake culture. This technique, which may combine spat collection with ongrowing, has not drastically changed since France's origin (13th century). Wooden poles are set into the intertidal seabed in rows (bouchots) (Figure 6). Each pole is 4-7 m long, 15-25 cm in diameter, and protrudes 2-3 m above the seabed. Several types of wood are currently used, including pine, oak, and, more recently, Brazilian squared hardwood. The rearing structures are 50-60 m long, with 120-130 poles in single or double lines for spat settlement and 80-90 poles for ongrowing. Bouchots should be spaced 25 m apart. Spat settlement occurs intensively in spring directly onto the wooden poles or the horizontal coconut fiber ropes strung on the poles before settlement. The seed is then transferred to tubular nets reattached around the growing poles in summer. The mesh tubes are nailed on at both ends.

Eventually, mussel seed spreads to cover the entire pole. Each pole produces around 60 kg live weight of mussels (Barnabe, 1994; FAO, 2009).

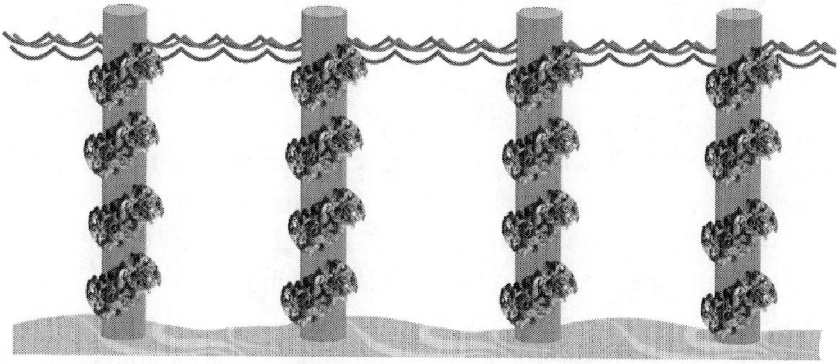

Figure 6. Bouchot culture.

The climate impacts on bouchot culture are similar to the rack method and horizontal culture. During turbulent weather, the strength of wooden poles can be damaged by strong wind and waves. Intense rainfalls result in the reduction of seawater salinity and contamination of marine pollutants; subsequently, physiological change in green mussels. High air temperature can cause thermal stress in green mussels during low tide. However, the bouchot method may receive smaller impacts. The structure of bouchot culture is larger and stronger than other methods; hence, it can be more resistant to turbulent weather. In addition, this method can locate in deep areas (>4 m) where are fewer climate stressors than in shallow waters. The cultivation areas should locate far from the river mouth or culture in bay with steady wind and waves to mitigate climate effects. Then, this cultivation method can be applicable for mussel culture under climate change crises.

Seed Collection

Green mussel spats are known as mussel seeds for mussel cultivation. Seeds are likely dispersed 10 km or less from shore (Gilg, et al., 2014). The site selected for seed collection should be free from pollutants. The ideal size of the seed is 15-25 mm in length or 1-2 g weight. All seeds are naturally attached with suspended rope or cotton net. The length of the seeded rope must be above the ground and tied so that the upper seeded portion of the rope should not get exposed during the low tide. The seeds will secrete byssus thread and attach themselves to the rope. Knots are made on the seeded rope at a distance of 25 cm to avoid slipping of the mussels. Placing split bamboo pegs in the rope (12-14 mm) at regular intervals will also serve the purpose (Alloycious, 2016).

Settlement or seeding patterns depend on water flow, oxygen concentration, temperature, and food availability. The settlement of green mussels increases with increasing water flow (Alfaro, 2005). The average velocity of seawater flowing passing the area of mussel cultures which have the potential of mussel settlement is in the range of 0.2-0.4 m/sec (Anongponyoshun and Meksumpun, 2007). High oxygen concentration can enhance the settlement of mussel larvae. Research by Alfaro (2005) revealed that mussels settlement increased when dissolved oxygen (DO) increased from 6 to 12 mg/L. Elevating seawater temperature and chlorophyll-*a* concentration can enhance green mussel spawning; subsequently, more mussel seeds. Research in the reproductive strategy of green mussel by MacFarland et al. (2014) releaved that mussels could spawn throughout the year. The average number of eggs released per female in September (summer) ($6.4 \times 10^6 \pm 2.6 \times 10^6$) was significantly higher than that in January (winter) ($7.7 \times 10^4 \pm 1.4 \times 10^4$). They also found that mussel's condition index increased with the rise in glycogen and lipid levels which coincided with increasing temperature and the chlorophyll-*a* level. It demonstrates that green mussel seeds will increase when the food supply is high.

Environmental variability caused by climate change can significantly affect the growth and reproduction of green mussels. During extreme weather events, floods, tremendous river runoff, and eutrophication result in salinity stress, hypoxia, and acidification in the aquatic environment (Hader and Barnes, 2019). These will affect all aquatic organisms, including green mussels. The decline of salinity and dissolved oxygen can cause high mortality of mussel larvae (Alfaro, 2005). Low pH levels in seawater can inhibit calcification which is a process to build a mussel's shell. Even though seawater circulation can support dispersion of mussel spats and enhancement of mussel seeding, strong wind and waves during turbulent weather can detach mussel seeds from the cultivation rope. There is not much research on this issue at present, and it may be a site-specific effect. Therefore, the response of green mussel seeding patterns to the new environment resulting from climate change still requires more investigation.

Growth Phase

Green mussels are generally grown in the open sea, coastal and estuarine areas. When mussel seeds are attached to ropes, they show fast growth in the suspended column. The crowded portion can slip weak mussels. To avoid slipping, the thinning of the same mussel size is essential. The ropes also should be suspended and not touch the bottom. The seeded portion should not be exposed for a long period during low tide. Seeded mussel on the upper portion of the rope shows faster growth due to the abundance of phytoplankton. For better water circulation, which brings enough food for mussel growth, the seeded ropes should be spaced at 25 cm. In open sea and coastal farming, the growth of mussels is very rapid. They attain 80-110 mm in 5-6 months with an average growth of 13.5 mm/month and an average weight of 35-45 g. In estuarine farming, mussels attain 75-90 mm in 5 months with an average weight of 35-40 g and 10 -12 kg/m rope production. The average growth rate is 1.13 g/month (Rajagopal et al., 1998a). Constant vigil is required to see that the raft/rack is positioned. Thinning process may be necessary to avoid loss of mussels and provide enough growing space. Periodic removal of fouling organisms such as barnacles, polychaetes, and ascidians should improve mussel growth (Alloycious, 2016).

During climate change events, variation in environmental conditions (strong currents and storms) can change the physiological response of green mussels. It was projected that at the end of the century (the year 2100), global atmosphere temperature would increase 1.0-5.7 °C (IPCC, 2021). This phenomenon causes high seawater temperature and influences mussel health (Morrill at a., 2001; Srisunont et al., 2022). Extreme rainfall events during climate change can cause salinity decreases due to high freshwater discharge from the river (Alosairi et al., 2019). Furthermore, it causes tremendous soil erosion; subsequently, high concentration of suspended sediments and dissolved organic matter in the water column (Pilditch et al., 2008). In addition, rainwater added into surface water and discharged through river runoff can reduce pH levels in seawater. Then, these changes in seawater parameters will decrease filtration rate, respiratory rate, hemocyte production, immune function, osmoregulation, and condition index in green mussels (Sasikumar and Krishnakumar, 2011; Wang et al., 2012). Finally, large mussel mortality can be found.

Diseases and Poisoning by Green Mussels Consumption

Green mussels have harmful effects when exposed to toxic phytoplankton, diseases, bacteria, or heavy metals in seawater. When marine pollutions occur, mussels consume all pollutants through their filtration process and accumulate in their bodies. Once polluted mussels are consumed, they cause disease and poisoning in consumers, including humans.

Heavy rainfalls cause drainage or runoff water from the surrounding area connecting the river during climate change events. Consequently, there is pollution in the receiving seawater body, especially in estuarine and coastal areas (Xu et al., 2019). Storms result in nutrient enrichment in the water column by direct decomposition and horizontal advection of nutrients from lands (Valiela et al., 2012). It can enhance phytoplankton (algae blooms) in estuarine and coastal areas known as the red tide (Alosairi et al., 2019; Meng et al., 2017). This phenomenon significantly impacts green mussel growth and survival (Noor et al., 2021). For instance, the red tide of dinoflagellate *Karenia brevis* (plankton density over 10^5 cells/mL) caused a growth rate reduction in mussels, and large mussel mortality was observed (McFarland et al., 2016). Once there is toxic phytoplankton bloom, these toxicants will accumulate in the mussel body and cause shellfish poisoning disease. Mussels will produce potent toxins that can find their way through the food chain to humans, causing various gastro-intestinal and neurological illnesses, such as paralytic shellfish poisoning, diarrhetic shellfish poisoning, amnesic shellfish poisoning, neurotoxic shellfish poisoning (Alloycious, 2016). Not only phytoplankton, but the magnitude of fecal coliform bacteria also increases with increasing rainfall intensity (Coulliette and Noble, 2008). These bacteria cause waterborne diseases that can disperse and contaminate aquatic organisms, including green mussels. Another pollutant is heavy metals. Heavy metal discharge through river runoff during intense rainfall can cause a physiological change in green mussels. Heavy metal can damage such exposure, often causing gill damage or closing mussel valves. Mussels may modify their behavior to cope with environmental stress by increasing biodeposition (Martinez et al., 2019); however, some metal particles accumulate in the mussel body. It seems that green mussels can contaminate with disease and toxicants during climate change events. The mussel farm should operate far from shore or less influence of pollutant and river discharge to mitigate this effect.

Conclusion

During climate change events, the green mussel population structure can be influenced by seawater parameters due to rising air temperature and intense rainfalls. Increased air temperature cause increasing seawater temperature; subsequently, thermal stress in mussels. Heavy rainfalls can reduce seawater salinity and pH level, enhancing the loading of suspended solids, organic matters, and nutrients. These changes impact mussels' physiological response and immune function. Consequently, mass mussel mortality can occur. In mussel cultivation, climate change events can affect mussel production. During extreme weather turbulence, the structural strength of the mussel farm can be destroyed by strong wind and waves. Mussel spawning and growth rate can be declined by rapid salinity reduction, hypoxia, and low pH level. Finally, tremendous pollutants discharge and red tide can cause disease and poisoning in mussels. There is still little knowledge of climate mitigation on mussel cultivation, which can be a site-specific impact. Therefore, mussel responds to climate change, and appropriate farm operation should be further investigated under climate change events.

References

Alfaro, A. C., Jeffs, A. G. and Hooker, S. H. (2001). Reproductive behavior of the green-lipped mussel, *Perna canaliculus*, in northern New Zealand. *Bulletin of Marine Science* 69(3), 1095-1108.

Alfaro, A. C., McArdle, B. and Jeffs, A. G. (2010). Temporal patterns of arrival of beachcast green-lipped mussel (*Perna canaliculus*) spat harvested for aquaculture in New Zealand and its relationship with hydrodynamic and meteorological conditions. *Aquaculture* 302, 208–218.

Alfaro, C. (2005). Effect of water flow and oxygen concentration on early settlement of New Zealand green-lipped mussel, *Perna canaliculus*. *Aquaculture* 246 (1-4), 285-294.

Alloycious, P. S. (2016). *Mussel Culture*. In: *Training Manual on Theeranaipunya - Equipping Fisherwomen Youth for Future*. ICAR-Central Marine Fisheries Research Institute, Kochi, 183-190. http://eprints.cmfri.org.in/10809/1/ Theeranaipunya_ Alloycious.pdf.

Alosairi, Y., Alsulaiman, N. and Karam Q. (2019). Responses of salinity and chlorophyll-a to extreme rainfall events in the northwest Arabian Gulf: Emphasis on Shatt Al-Arab. *Marine Pollution Bulletin* 149, 1-7.

Anongponyoshun, M. and Meksumpun, S. (2007). Increasing area of green mussel culture in Ao Sriracha, Chon-Buri Province. *Kasetsart Journal (Natural Science)* 41, 690-695.

Barnabe, G. (1994). *Aquaculture: biology and ecology of cultured species*. Ellis Horwood *Series in Aquaculture and Fisheries Support*. UK: Wiley & Sons.

Cebu, E. H. and Orale, R. L. (2017). Seawater physicochemical parameters in the green mussel belts in Samar Philippines. *Journal of Academic Research* 02:4. 1-15.

Cheung, S. G. and Shin, P. K. S. (2005). Size effects of suspended particles on gill damage in green-lipped mussel *Perna viridis*. *Marine Pollution Bulletin* 51, 801-810.

Coulliette, A. D. and Noble, R. T. (2008). Impacts of rainfall on the water quality of the Newport River Estuary (Eastern North Carolina, USA). *Journal of Water and Health* 06.4, 473-482.

Donaghy, L., Volety, A. K. 2011. Functional and metabolic characterization of hemocytes of the green mussel, *Perna viridis*: *in vitro* impacts of temperature. *Fish and Shellfish Immunology* 31, 808-814.

Doney, S. C., Ruckelshaus, M., Duffy, J. E., Barry, J. P., Chan, F., English, C. A., Galindo, H. M., Grebmeier, J. M., Hollowed, A. B., Knowlton, N., Polovina, J., Rabalais, N. N., Sydeman, W. J., and Talley, L. D. (2012). Climate change impacts on marine ecosystems. *Annual Review of Marine Science* 4, 11–37.

Doney, S. C., Ruckelshaus, M., Duffy, J. E., Barry, J. P., Chan, F., English, C. A., Galindo, H. M., Grebmeier, J. M., Hollowed, A. B., Knowlton, N., Polovina, J., Rabalais, N. N., Sydeman, W. J. and Talley, L. D. (2012). Climate change impacts on marine ecosystems. *Annual Review of Marine Science* 4, 11–37.

Dore, I. (1991). *Shellfish: a guide to oyster, mussels, scallops, clams and similar products for the commercial user*. New York, USA: Van Nostrand Reinhold.

Duarte, C., Navarro, J. M., Acuña, K., Torres, R., Manríquez, P. H., Lardies, M. A., Vargas, C. A., Lagos, N. A., and Aguilera, V. (2014). Combined effects of temperature and ocean acidification on the juvenile individuals of the mussel *Mytilus chilensis*. *Journal of Sea Research* 85, 308–314.

Duarte, C., Navarro, J. M., Acuña, K., Torres, R., Manríquez, P. H., Lardies, M. A., Vargas, C. A., Lagos, N. A. and Aguilera, V. (2014). Combined effects of temperature and ocean acidification on the juvenile individuals of the mussel *Mytilus chilensis*. *Journal of Sea Research* 85, 308–314.

Eglash, R. E. (2005). Effect of Rainfall on Dissolved Oxygen in the Palo Alto Flood Control Basin. *California State Science Fair 2005 Project Summary*, 05.

European Environment Agency. (2021). *Ocean Acidification*. Retrieved Jan 25, 2022 from https://www.eea.europa.eu/ims/ocean-acidification.

Fang, J. K. H., Wu, R. S. S., Chan, A. K. Y., Yip, C. K. M. and Shin, P. K. S. (2008). Influences of ammonia-nitrogen and dissolved oxygen on lysosomal integrity in green-lipped mussel *Perna viridis*: laboratory evaluation and field validation in Victoria Harbour, Hong Kong. *Marine Pollution Bulletin* 56, 2052-2058.

FAO. (2009). *Mytilus edulis*. In Goulletquer, P. (Ed.). *Cultured aquatic species fact sheets*. CD-ROM (multilingual).

FAO. (2021). *FAO Yearbook. Fishery and Aquaculture Statistics 2019/FAO annuaire*. Rome. 110 p.

Firth, L. B., Knights, A. M. and Bell, S. S. (2011). Air temperature and winter mortality: Implications for the persistence of the invasive mussel, *Perna viridis* in the intertidal zone of the south-eastern United States. *Journal of Experimental Marine Biology and Ecology* 400(1-2), 250-256.

Fitzer, S. C., Cusack, M., Phoenix, V. R. and Kamenos, N. A. (2014). Ocean acidification reduces the crystallographic control in juvenile mussel shells. *Journal of Structural Biology* 188, 39-45.

Gao, Q. F., Shin, P. K. S., Lin, G. H., Chen, S. P., and Cheung, S. G. (2006). Stable isotope and fatty acid evidence for uptake of organic waste by green-lipped mussels *Perna viridis* in a polyculture fish farm system. *Marine Ecology Progress Series* 317, 273-283.

Gao, Q. F., Xu, W. Z., Liu, X. S, Cheung, S. G. and Shin, P. K. S. (2008). Seasonal changes in C, N and P budgets of green-lipped mussels *Perna viridis* and removal of nutrients from fish farming in Hong Kong. *Marine Ecology Progress Series* 353, 137-146.

Gestoso, I., Arenas, F. and Olabarria, C. (2016). Ecological interactions modulate responses of two intertidal mussel species to changes in temperature and pH. *Journal of Experimental Marine Biology and Ecology* 474, 116–125.

Gestoso, I., Arenas, F., and Olabarria, C. (2016). Ecological interactions modulate responses of two intertidal mussel species to changes in temperature and pH. *Journal of Experimental Marine Biology and Ecology* 474, 116–125.

Gilg, M. R., Howard, R., Turner, R., Middlebrook, M., Abdulnour, M., Lukaj, E., Sheng, Y. P., Liu, T. and Tutak, B. (2014). Estimating the dispersal capacity of the introduced green mussel, *Perna viridis* (Linnaeus, 1758), from field collections and oceanographic modeling. *Journal of Experimental Marine Biology and Ecology* 461, 233-242.

Goh, B. P. L. and Lai, C. H. (2014). Establishing the thermal threshold of the tropical mussel *Perna viridis* in the face of global warming. *Marine Pollution Bulletin* 85, 325–331.

Goh, B. P. L. and Lai, C. H. (2014). Establishing the thermal threshold of the tropical mussel *Perna viridis* in the face of global warming. *Marine Pollution Bulletin* 85, 325–331.

Gu, H., Shanga, Y., Clements, J., Dupont, S., Wang, T., Wei, S., Wang, X., Chen, J., Huangg, W., Hu, M. and Wang, Y. (2019). Hypoxia aggravates the effects of ocean acidification on the physiological energetics of the blue mussel *Mytilus edulis*. *Marine Pollution Bulletin* 149, 1-7.

Hader, D. P. and Barnes, P. W. (2019). Comparing the impacts of climate change on the responses and linkages between terrestrial and aquatic ecosystems. *Science of Total Environment* 682, 239-246.

Harvey, R., Lye, L., Khan, A. and Paterson, R. (2013). The Influence of Air Temperature on Water Temperature and the Concentration of Dissolved Oxygen in Newfoundland Rivers. *Canadian Water Resources Journal* 36(2), 171–192.

Hawkins, A. J. S., Bayne, B. L., Bougrier, S., Heral, M., Iglesias, J. I. P, Navarro, E., Smith, R. F. M. and Urrutia, M. B. (1998). Some general relationships in comparing the feeding physiology of suspension feeding bivalve molluscs. *Journal of Experiment Marine Biology and Ecology* 219, 87-103.

Hawkins, A. J. S., James, Hickman, R. W., Hatton, S. and Weatherhead, M. (1999). Modelling of suspension-feeding and growth in the green-lipped mussel *Perna canaliculus* exposed to natural and experimental variations of seston availability in the Marlborough Sounds, New Zealand. *Marine Ecology Progress Series* 191, 217-232.

Hickman, R. W. (1979). Allometry and growth of the green-lipped mussel *Perna canaliculus* in New Zealand. *Marine Biology* 51(4), 311-327.

Hickman, R. W., Waite, R. P, Illingworth, J., Meredyth-Young, J. L. and Payne, G. (1991). The relationship between farmed mussels, *Perna canaliculus,* and available food in Pelous-Kenepuru Sound, New Zealand, 1983-1985. *Aquaculture* 99, 49-68.

Huhn, M., Hattich, G. S. I., Zamani, N. P., von Juterzenka, K. and Lenz, M. (2016). Tolerance to stress differs between Asian green mussels *Perna viridis* from the impacted Jakarta Bay and from natural habitats along the coast of West Java. *Marine Pollution Bulletin* 110(2), 757-766.

Information and communication technology center. (2021). *Fisheries Statistic of Thailand 2019*. Department of Fisheries Technical Paper No. 2/2021.

IPCC. (2021). Summary for Policymakers. In: Climate Change 2021: The Physical Science Basis. Contribution of Working Group I to the Sixth *Assessment Report of the Intergovernmental Panel on Climate Change* [Masson-Delmotte, V., P. Zhai, A. Pirani, S. L. Connors, C. Péan, S. Berger, N. Caud, Y. Chen, L. Goldfarb, M. I. Gomis, M. Huang, K. Leitzell, E. Lonnoy, J. B. R. Matthews, T .K. Maycock, T. Waterfield, O. Yelekçi, R. Yu, and B. Zhou (eds.)]. In Press.

Johnson, F. (1971). Stream temperatures in an alpine area. *Journal of Hydrology* 14, 322-336.

Lee, S. Y. (1985). The population dynamics of the green mussel *Perna viridis* (L.) in Victoria Harbour, Hong Kong - Dominance in a polluted environment. *Asian Marine Biology* 2, 107-118.

Liutkus, M., Robinson, S., MacDonald, B., and Reid, G. (2012). Quantifying the effects of diet and mussel size on the biophysical properties of the blue mussel, *Mytilus* spp., feces egested under simulated imta conditions. *Journal of Shellfish Research* 31(1), 69-77.

MacFarland, K., Soudant, P., Jean, F. and Volety, A. K. (2014). Reproductive strategy of the invesive green mussel may result in increased competition with native fauna in the southeastern United States. *Aquatic Invastions* 11(4), 411-423.

Martinez, A. S., Mayer-Pinto, M. and Christofoletti, R. A. (2019). Functional responses of filter feeders increase with elevated metal contamination: Are these good or bad signs of environmental health? *Marine Pollution Bulletin* 149, 110571.

McFarland, K., Donaghy, L. and Volety, A. K. (2013). Effect of acute salinity changes on hemolymph osmolality and clearance rate of the non-native mussel, *Perna viridis*, and the native oyster, *Crassostrea virginica*, in Southwest Florida. *Aquatic Invastions* 8(3), 299-310.

McFarland, K., Jean, F., Thébault, J. and Volety, A. K. (2016). Potential impacts of blooms of the toxic dinoflagellate *Karenia brevis* on the growth, survival and juvenile recruitment of the non-native green mussel *Perna viridis* in southeastern United States. *Toxicon* 109, 94-102.

Meng, P. J., Tew, K. S., Hsieh, H. Y., Chen, C. C. (2017). Relationship between magnitude of phytoplankton blooms and rainfall in a hyper-eutrophic lagoon: A continuous monitoring approach. *Marine Pollution Bulletin* 124, 897-902.

Morrill, J. C., Bales, R. C. and Conklin, M. H. (2001). The Relationship Between Air Temperature and Stream Temperature. *American Geophysical Union, Spring Meeting 2001*, abstract id. H42A-09.

Nair R, M. and Appukuttan, K. K. (2003). Effect of temperature on the development, growth, survival and settlement of green mussel *Perna viridis* (Linnaeus, 1758). *Aquaculture Research* 34, 1037-1045.

Noor, A. R., Shakil, A., Hoque, N. F., Rahman, M. M., Akter, S., Talukder, A., Ahmad-Al-Nahid, S., Wahab, M. A., Nahiduzzaman, M., Rahman, M. J. and Asaduzzaman, M. (2021). Effect of eco-physiological factors on biometric traits of green mussel Perna viridis cultured in the south-east coast of the Bay of Bengal, Bangladesh. *Aquacullture Reports* 19, 100562.

Northeastern Regional Aquaculture Center (NRAC). (1993). Major Predators of Cultured Shellfish. *NRAC Bulletin* No. 180.

Pernet, F., Malet, N., Pastoureaud, A., Vaquer, A., Quere, C., and Dubroca, L. (2012). Marine diatoms sustain growth of bivalves in a Mediterranean lagoon. *Journal Sea Research* 68, 20-32.

Pilditch, C. A., Widdows, J., Kuhn, N. J., Pope, N. D. and Brinsley, M. D. (2008). Effects of low tide rainfall on the erodibility of intertidal cohesive sediments. *Continental Shelf Research* 28(14), 1854-1865.

Rajagopal, S., Venugopalan, V. P., Nair, K. V. K., van der Velde, G., and Jenner, H. A. (1998b). Settlement and growth of the green mussel *Perna viridis* (L.) in coastal water: influence of water velocity. *Aquatic Ecology* 32, 313-322.

Rajagopal, S., Venugopalan, V. P., Nair, K. V. K., van der Velde, G., Jenner, H. A. and den Hartog. (1998a). Reproduction, growth rate and culture potential of the green mussel, *Perna viridis* (L.), in Edaiyur backwaters, east coast of India. *Aquaculture* 162, 187-202.

Rajagopal, S., Venugopalan, V. P., van der Velde, G., and Jenner, H. A. (2006). Greening of the coasts; a review of the *Perna viridis* success story. *Aquatic Ecology* 40, 273-297.

Rajesh, K. V., Mohamed, K. S., and Kripa, V. (2001). Influence of algal cell concentration, salinity and body size on the filtration and ingestion rates of cultivable India bivalves. *Indian Journal of Marine Sciences* 30, 87-92.

Reddy, N. D. and Menon, N. R. (1979). Effects of Ammonia and Ammonium on tolerance and Byssogenesis in *Perna viridis*. *Marine Ecology Progress Series* 1(4), 315-321.

Roungjai, S. (1999). *Effects of low salinity on mortality of green-lipped mussel* (*Perna viridis Linneaus*). Bangkok, Thailand; Faculty of Fisheries, Kasetsart University.

Sanders, T., Schmittmann, L., Nascimento-Schulze, J. C. and Melzner, F. (2018). High calification costs limit mussel growth as low salinity. *Frontia Marrine Science* 5, 352.

Sasikumar, G. and Krishnakumar, P. K. (2011). Aquaculture planning for suspended bivalve farming systems: The integration of physiological response of green mussel with environmental variability in site selection. *Ecological Indicators* 11, 734–740.

Shin, P. K. S., Yau, F. N., Chow, S. H., Tai, K. K. and Cheung, S. G. (2002). Responses of the green-lipped mussel *Perna viridis* (L.) to suspended solids. *Marine Pollution Bulletin* 45, 157-162.

Sivalingam, P. M. (1977). Aquaculture of the green mussel, *Mytilus viridis* Linnaeus, in Malaysia. *Aquaculture* 11(4), 297-312.

Sivalingam, P. M. (1977). Aquaculture of the green mussel, *Mytilus viridis* Linnaeus, in Malaysia. *Aquaculture* 11(4), 297-312.

Song, C., Pabst, A. and Bowers, C., (1973). Stochastic Analysis of Air and Water Temperatures. *Journal of Environmental Engineering* 99, 785-800.

Soon, T. K. and Ransangan, J. (2016). Feasibility of green mussel, *Perna viridis* farming in Marudu Bay, Malasia. *Aquaculture Reports* 4, 130-135.

Sreedevi, P. R., Uthayakumar, V., Jayakumar, R. and Ramasubramania, V. (2014). Influence of rearing water temperature on induced gonadal development and spawning behaviour of tropical green mussel, *Perna viridis*. Asian *Pacific Journal Reprodution* 3(3), 201-209.

Srisunont, C., Srisunont, T., Intarachart, A. And Babel, S. (2022). Influence of Climate on Seawater Quality and Green Mussel Production. *Scientia Marina* 86(1) https://doi.org/10.3989/scimar. 05232.027.

Sui, Y., Kong, H., Huang, X., Dupont, S., Hu, M., Storch, D., Portner, H. O., Lu, W. and Wang, Y. (2016a). Combined effects of short-term exposure to elevated CO_2 and decreased O_2 on the physiology and energy budget of the thick shell mussel *Mytilus coruscus*. *Chemosphere* 155, 207-216.

Sui, Y., Kong, H., Shang, Y., Huang, X., Wu, F., Hu, M., Lin, D., Lu, W. and Wang, Y. (2016b). Effects of short-term hypoxia and seawater acidification on hemocyte responses of the mussel *Mytilus coruscus*. *Marine Pollution Bulletin* 108, 46–52.

Tamtin, M., Sanggorntanagit, T. and Khunna, A. (2004). Feeding efficiency of green mussels (*Perna viridis* Linn., 1758) on water from shrimp culture pond. *Coastal Aquatic Feed Research Institute, Chon Buri, Thailand Technical Paper* 64.

Tantanasarit, C. (2013). *Effect of green mussels (Perna viridis) on nutrient dynamics in estuarine and coastal areas*. Doctoral dissertation, Sirinhorn International Institutye of Technology, Thammasat University, Thailand.

Tantanasarit, C., Babel, S., Englande, A. J., and Meksumpun, S. (2013). Influence of size and density on filtration rate modeling and nutrient uptake by green mussel (Perna viridis). *Marine Pollution Bulletin* 68, 38-45.

Trottet, A., Roy, S., Tamigneaux, E. and Lovejoy, C. (2007). Importance of heterotrophic planktonic communities in a mussel culture environment: the Grande Entrée lagoon, Magdalen Island (Quebes, Canada). *Marine Biology* 151, 377-392.

Tuaycharoen, S., Vakily, J. M., Sealow, A., and Mccoy, E. W. (1998). Growth and maturation of the green mussel (*Perna viridis*) in Thailand. Maila: Bivalve mollusk culture research in Thailand, *ICLARM Technical Reports* 19, 88-101.

Tunkijjanukij, S. and Intarachart, A. (2007). Development of green mussel cultivation in Thailand: Sriracha Bay, Chonburi Province. *Aquaculture Asia Magazine* 10(2), 24-25.

Vakily, J. M. (1989). The Biological and Culture of Mussels of the Genus *Perna. Manila: ICLARM Studies and Reviews* 17.

Valiela, I., Camilli, L., Stone, T., Giblin, A., Crusius, J., Fox, S., Barth-Jensen, C., Monteiro, R. O., Tucker, J., Martinetto, P. and Harris, C. (2012). Increased rainfall remarkably freshens estuarine and coastal waters on the Pacific coast of Panama: Magnitude and likely effects on upwelling and nutrient supply. *Global and Planetary Change* 92–93, 130-137.

van Khoi, L., and Fotedar, R. (2012). Integration of blue mussel (*Mytilus edulis* Linnaeus, 1758) with western king prawn (*Penaeus latisulcatus* Kishinouye, 1896) in a closed recirculating aquaculture system under laboratory conditions. *Aquaculture* 345-355, 84-90.

Wang, Y., Hu, M., Cheung, S. G., Shin, P. K. S., Lu, W. and Li, J. (2012). Immune parameter changes of hemocytes in green-lipped mussel *Perna viridis* exposure to hypoxia and hyposalinity. *Aquaculture* 356–357, 22–29.

Wang, Y., Hu, M., Cheung, S. G., Shin, P. K. S., Lu, W., and Li, J. (2012). Immune parameter changes of hemocytes in green-lipped mussel *Perna viridis* exposure to hypoxia and hyposalinity. *Aquaculture* 356–357, 22–29.

Wang, Y., Hu, M., Shin, P. K. S. and Cheung, S. G. (2009). Introduction of anti-predator responses in the green-lipped mussel *Perna viridis* under hypoxia. *Marine Biology* 157(4), 747-754 (https://doi.org/10.1007/s00227-009-1358-z).

Wang, Y., Hu, M., Shin, P. S. K. and Cheung, S. G. (2011). Immune responses to combined effect of hypoxia and high temperature in the green-lipped mussel *Perna viridis*, *Marine Pollution Bulletin* 63, 201-208.

Wendling, C. C., Huhn, M., Ayu, N., Bachtiar, R., Juterzenka, K. and Lenz, M. (2013). Habitat degradation correlates with tolerance to climate-change related stressors in the green mussel *Perna viridis* from West Java, Indonesia. *Marine Pollution Bulletin* 71, 222–229.

Wendling, C. C., Huhn, M., Ayu, N., Bachtiar, R., Juterzenka, K., and Lenz, M. (2013). Habitat degradation correlates with tolerance to climate-change related stressors in the green mussel *Perna viridis* from West Java, Indonesia. *Marine Pollution Bulletin* 71, 222–229.

Wong, W. H. and Cheung, S. G. (1999). Feeding behavior of the mussel, *Perna viridis* (L.): responses to variation in seston quantity and quality. *Journal of Experimental Marine Biology and Ecology* 236, 191-207.

Wong, W. H. and Cheung, S. G. (2001a). Feeding rates and scope for growth the green mussel, *Perna viridis* (L.) and their relationship with food availability in Kat O, Hong Kong. *Aquaculture* 193, 123-137.

Wong, W. H. and Cheung, S. G. (2001b). Feeding rhythms of the green-lipped mussel, *Perna viridis* (Linnaeus, 1758) (Bivalvia: Mytilidae) during spring and neap tidal cycles. *Journal of Experimental Marine Biology and Ecology* 257, 13-36.

Wong, W. H. and Cheung, S. G. (2003). Seasonal variation in the feeding physiology and scope for growth of green mussels, *Perna viridis* in estuarine Ma Wan, Hong Kong. *Journal of the Marine Biological Association of the United Kingdom* 83(3), 543-552.

Wu, F., Lu, W., Shang, Y., Kong, H., Li, L., Sui, Y., Hu, M. and Wang, Y. (2016). Combined effects of seawater acidification and high temperature on hemocyte parameters in the thick shell mussel *Mytilus coruscus*. *Fish & Shellfish Immunology* 56, 554-562.

Wu, Rudolf S. S. and Lam, P. K. S. (1997). Glucose-6-phosphate dehydrogenase and lactate dehydrogenese in the green-lipped mussel (*Perna viridis*): possible biomarkers for hypoxia in the marine environment. *Water Research* 31(11), 2797-2801.

Xu, H., Zhang, Y., Zhu, X. and Zheng, M. (2019). Effects of rainfall-runoff pollution on eutrophication in coastal zone: A case study in Shenzhen Bay, southern China. *Hydrology Research* 50.4, 1062-1074.

Xu, X. Y., Yip, K. R., Shin, P. K. S. and Cheung, S. G. (2017). Predator–prey interaction between muricid gastropods and mussels under ocean acidification. *Marine Pollution Bulletin* (http://dx.doi.org/10.1016/j.marpolbul.2017.01.003).

Biographical Sketch

Chayarat Srisunont

Affiliation: Environmental Science and Technology Program, Faculty of Science and Technology, Bansomdejchaopraya Rajabhat University, Thailand

Education:

1995-2001	Level 1-12 at Satit Kasetsart University Laboratory School, Thailand, GPA 3.34
2001-2005	Undergraduate: BSc. (Fisheries) with 1st class honor from Major Field: Marine Science, Department of Marine Science, Faculty of Fisheries, Kasetsart University, Thailand, GPA 3.63
2005-2007	Graduated: MSc. (Marine Science) from Major Field: Marine Science, Department of Marine Science, Faculty of Fisheries, Kasetsart University, Thailand, GPA 3.97 Thesis Title: Relationships between Environmental Factors and Production Potential of Razor Clam (*Solen* spp.) at Don Hoi Lord, Samut Songkhram Province.
2009-2012	Graduated: PhD in Technology from Environmental Technology program, School of Bio-Chemical Engineering and Technology, Sirindhorn International Institute of Technology (SIIT), Thammasat University, Thailand. Thesis Title: Effect of Green Mussels (*Perna viridis*) on Nutrient Dynamics in Estuarine and Coastal Areas.

Business Address: Faculty of Science and Technology, Bansomdejchaopraya Rajabhat University, 1061, Issaraphab Rd. Soi 15, Hirunruchi, Thonburi, Bangkok, 10600, Thailand

Research and Professional Experience:

2018-2020 *Project Leader* in research entitled of "Green Mussel Cultivation Model: A Climate Change Adaptation Tool for Smart Farmers" Financial support by Thailand Research Fund.

2016-2017 *Project Leader* in research entitled of "Influence of Environmental Factors on Blood Cockles Potential at Klong Khone, Samut Songkharm Province" Financial support by National Research Council of Thailand.

2015-2016 *Project Leader* in research entitled of "Influence of water quality on algae distribution in shrimp ponds at Bangkhuntein District, Bangkok" Financial support by Research and Development Institute, Bansomdejchaopraya Rajabhat University, Thailand.

2009-2012 *Project Leader* in research entitled of "Effect of Green Mussels (*Perna Viridis*) on Nutrient Dynamics in Estuarine and Coastal Areas" Award scholarship from The Royal Golden Jubilee Ph.D. Program (RGJ) under Thailand Research Fund (TRF).

Professional Appointments: Assistant Professor

Honors:

2015 Best Poster Presentation Award in "National Conference on Conservation Status, Present and Future of Biodiversity" at Trang, Thailand on June 10-12, 2015." Entitled of "Influence of environmental factors on blood cockle density at Klong Khone, Samut Songkhram Province"

2013 Outstanding Research Award that contributes to the Sirindhorn International Institute of Technology (SIIT) reputation on June 1, 2013. The research is entitled of "Effect of green mussels (*Perna viridis*) on nutrient dynamics pin estuarine and coastal areas" (Doctoral thesis academic year 2012).

2012 Outstanding Research Award in "RGJ-Ph.D. Congress XIII" by The Thailand Research Fund (TRF) (April 6-8, 2012), entitled of "Investigation on Filtration Rate and Subsequent Nutrient Removal by Green Mussel (*Perna viridis*)". This award was given to only 5 researches which were selected from 275 presented researches in the congress. This research was chosen for media interview by TRF committee, and was published in "U-Challenge" magazine for young education (university and high school students) year 6 Vol. 67 May 2012, page 28-30. This received award also was published in "SIIT News", Thammasat University, Vol.17 No.2 June 2012, page 9.

2012 Outstanding Oral Presentation Award in "RGJ-Ph.D. Congress XIII" by The Thailand Research Fund (TRF) (April 6-8, 2012), entitled of "Investigation on Filtration Rate and Subsequent Nutrient Removal by Green Mussel (*Perna viridis*)".

Publications from the Last 3 Years:

1. Chayarat Srisunont, Yokrudee Nobpakhuna, Chontara Yamaleea, Treeranut Srisunontb, (2018). Seasonal variation in seawater quality and its correlation with blood cockle at Klong Khone, Thailand. *Proceedings of International Conference of Agriculture and Natural Resources (ANRES 2018),* 26th – 28th April 2018, pp. 268-271.
2. Treeranut Srisunont, Chayarat Srisunont and Rattachon Angmanee. (2018). Applying STELLA Software for Streamflow Modeling in U-Tapao Watershed. Proceeding *The 7th International Conference on Environmental Engineering, Science and Management,* May 24-25, 2018, pp. 61-62.
3. Treeranut Srisunont, Sittichai Tantanasarit and Chayarat Srisunont. (2019). Applying STELLA model to optimize land allocation in watershed based on DO and BOD dynamics. *Suan Sunandha Science and Technology Journal,* 5(2), pp. 1-15.
4. Chayarat Srisunont, Yokrudee Nobpakhun, Chontara Yamalee and Treeranut Srisunont. (2020). Influence of Seasonal Variation and Anthropogenic Stress on Blood Cockle (*Tegillarca granosa*) Production Potential. *Journal of Fisheries and Environment,* 44 (2), pp. 62-82.
5. Treeranut Srisunont and Chayarat Srisunont. (2022). Water Quality and Sediment Quality of Cockle Culture Area in the Upper Gulf of Thailand. *Current Applied Science and Technology*, 22(3), pp. 1-20. (https://doi.org/10.55003/cast.2022.03.22.003).
6. Treeranut Srisunont and Chayarat Srisunont. (2022). Influence of Environmental Factors on Blood Cockle Production Potential at Klong Khone, Samut Songkharm Province and Bang Taboon, Phetchaburi Province. *Current Applied Science and Technology*, 22(4), pp. 1-21. (https://doi.org/10.55003/cast.2022.04.22.008).

7. Chayarat Srisunont, Treeranut Srisunont, Alongot Intarachart and Sandhya Babel. (In Press). Influence of climate on seawater quality and green mussel production. *Scientia Marina*, 86(1), pp. 1-22. (https://doi.org/10.3989/scimar.05232.027).

Chapter 3

Cetaceans in the Tropical Savu Sea, Indonesia

Agus Hartoko*
Jahved Ferianto Maro
and Delianis Pringgenies
Faculty of Fisheries and Marine Science, Diponegoro University,
Semarang, Indonesia

Abstract

A holistic concept of research on marine mammals or cetaceans in a series of time and spatial oceanography database, eco-biology factors describing the diversity of cetacean records ie. appearance of stranded whales, apperance of dolphins, habitat of dugong at the very specific tropical Savu Sea as part of global scale migration tracks, which is presumably outside of their natural habitat in the polar region. Exploring how these specific oceanographical phenomenas or events affect local marine ecosystems and the cetacean communities, in an ever changing world based on the spatio-temporal oceanographic data-base record to provide a strong scientific basis for marine conservation decisions. Development of diverse research platforms enhancing traditional knowledge of many species of whales stranded, with an advanced cutting-edge technology such as DNA of cetaceans espescially the very often stranded whale and their natural feeding in the tropical Savu Sea. The management plan would be introduced as adaptive local wisdom and scientific based sustainable "cetacean-track" for eco-friendly tourism in new concept of "marine geopark."

Keywords: cetacean, Savu Sea, ecobiology, spatial-analysis, dolphin, dugong

* Corresponding Author's Email: agushartoko.undip@gmail.com.

In: Advances in Marine Biology. Volume 6
Editors: Adam Kovács and Patrik Nagy
ISBN: 979-8-88697-996-1

Dugong

Introduction

Dugong is the only sea ox that can be found in the waters of 37 countries including Indonesia. Most dugongs inhabit Eastern Indonesia and Northern Australia (Kiszka, 2007; Garrigue, Patenaude and Marsh, 2008; Blair et al., 2014). Indonesian waters have approximately 35 species of mammals, one being the dugong-dugon (Dewi et al., 2018). The habitat of dugong is commonly found around coastal or inshore waters that enable dugong to access shallow as well as deep water (Anderson, 1981). Dugong has now become a tourist attraction with an increasing number of visitors (Juraij, 2014; Nugraha et al., 2019). In Indonesia, there are several waters that support the existence of dugong (Dugong-dugon) such as the waters of Papua, Sulawesi, Lombok and Java (Priosambodo et al., 2007; Dewi et al., 2018; Marsh, Harris and Lawler, 1997; Karim, Kusuma and Mahfud, 2019; Lee and Nijman, 2015; Nurdin et al., 2019; Priosambodo et al., 2007).

Indonesian waters have many habitats for dugong (Dugong-dugon) but they do not have a friendly relationship with humans as do the dugongs in Kabola, Alor, Indonesia. Therefore, based on the background of the problem in this article, the authors would like to do specific research on the behavior of dugongs that interact actively with humans in the waters of Kabola Beach, Alor Island, Indonesia.

The waters of the Alor Strait and its surroundings are located in the Province of East Nusa Tenggara, which is one of the areas located in the Indonesian Exclusive Economic Zone next to the west coast. Timor Leste and Australia are also the areas of the trajectory of the Indonesian Throughflow, considered as the confluence of two current masses from the Pacific Ocean and Indian Ocean (Putra et al., 2016). The waters of the Alor Strait are unique with dynamic oceanographic variables such as a significant change in sea surface and vertical temperature variability and salinity during the southeast monsoon season. The dynamic of the waters occur in the surface layer which is influenced by monsoon wind patterns. This condition results in upwelling in the Savu Sea waters. The process of a stirred up water mass (upwelling) in the waters affects the living conditions of phytoplankton, hydrology, and the nutrient enrichment in the waters (Sediadi, 2010; Packard et al., 2015). One of the significant impacts of upwelling is the increasing fertility (abundance of plankton as natural food) and the increasing sea water temperature (warm), thus providing comfort for a group of cetaceans, including the humpback whale and the bottle-nose dolphin which are migrating in the waters of the Alor Strait (Mujiyanto, et al., 2018).

The Alor Strait is part of the Pantar Strait Conservation Area Nature Preserve and its surrounding sea which has been arranged in the Ministerial Regulation No. 5/KEPMEN-KP/2015 by the Ministry of Marine Affairs and Fisheries Republic of Indonesia. The appearance of the humpback whale in the Alor Strait waters was initially informed by some fishermen in the location. The fishermen reported the information to the researchers so further monitoring and research on this species could be conducted. Therefore based on the background of the problem, the monitoring research on the humpback whale in the Alor Strait waters using field observation and observed the bio-oceanography of the humpback whale method is inevitable.

Literature Review

Dugong (*Dugong-dugon*) is a species of marine mammal categorized in the order *Sirenia* or sea cow that still exists other than the manatee. In general dugong has life range of 22 to 25 years (Sheppard et al., 2010; Sheppard et al., 2006; Aragones et al., 2006; Marsh, Harris and Lawler, 1997). Dugong is not classified as a fish because it breastfeeds its baby called a calf and is categorized as the evolutionary relative of the elephant. Dugong is the only representative of family *Dugongidae* (Aketa et al., 2003; Hodgson, 2009; Grech, Sheppard and Marsh, 2011).

An on-site behavior study of dugong (*Dugong-dugon*) Sirenia: Dugongidae (Muller, 1775) was carried out at Kabola Beach, Alor Island. The research was conducted during east season from July to October, 2020. Field observation had found aggressive and familiar behavior of dugong to humans who visited the location. The study had identified the dugong at Kabola Beach as male with 2.6 meters of body length and 1.5 meters of body circumference that inhabits a seagrass bed coastal ecosystem. The species of seagrass mostly consumed by dugong are *Halophila ovalis* and *Enhalus acoroides*. The specific behavior of the dugong's aggressive interaction with humans are a rise to the surface for about 3 seconds, then he swims back down into the water for 4 to 5 minutes, then pops up again to the surface. This dugong sanctuary habitat is exclusively maintained and conserved by Mr Onesimus Laa.

Method

Field Observation

This research was conducted from July to October 2020 with field observation. During research, the authors documented every movement of dugong and its interaction with humans (Juraij, 2014). The seawater depth obtained from General Bathymetric Chart of the Oceans (GEBCO), www.gebco.net. The appearance of dugong (Dugong-dugon) in Kabola waters was at coordinates 8°7.553'S - 124°36.807'E. The map of the research location can be seen in Figure 1.

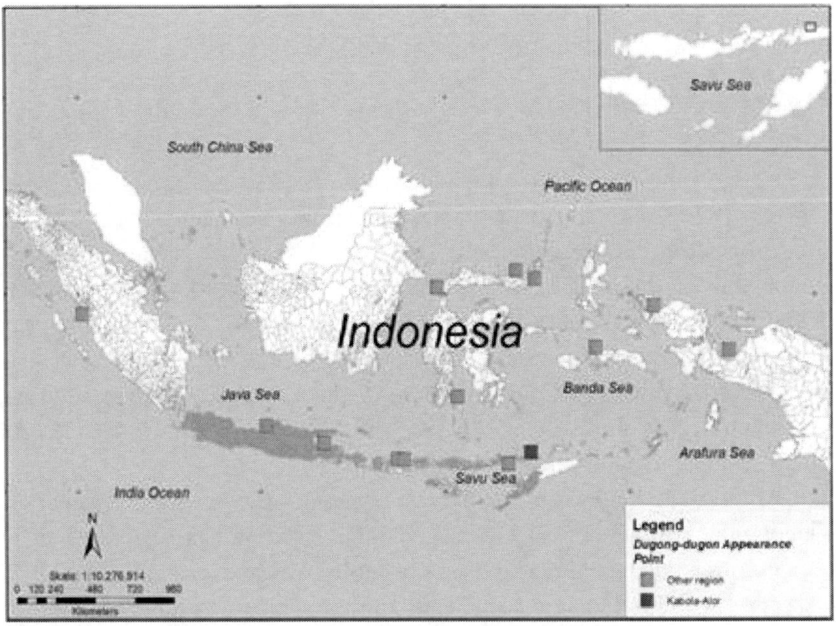

Figure 1. Dugong appearance and habitat points in Indonesia and at Kabola Beach, Alor.

Result and Discussions

The field observations revealed that the dugong in Kabola Beach Alor specifically had friendly behavior with the locals as well the tourists. The taxononimy of dugong (*Dugong-dugon*) are Family *Dugongidae*, Species *Dugong-dugon.* Dugong are locally called 'duyung' which is closely relative to steller sea cow (*Hydrodamalis gigas*). Most of *H. gigas* itself became extinct presumably due to decreasing coastal water quality and seagrass biomass. Commonly, dugong has a large head, dark pupils and pale white eyeballs. The body is bluish gray, pale to pink on the lower body. Adult male dugong usually have scars on its body. Dugong has a paddle-shaped body and its tail (flukes) has a notch in the center like a whale's tail (Muller, 1766 dalam (Evans, 2008; Hodgson, 2009). The visual morphology of dugong (*Dugong-dugon*) in Kabola coastal water is presented in Figure 2.

The length and weight measurements of dugong (*Dugong dugon*) were identified by digitizing pictures taken during research. The pictures consist of a male dugong with 2.6 meters of body length, 1.5 meters of body circumference, 25 centimeters of tail width, and 17 centimeters of flipper length. According to James (1974); Aketa et al., (2003); Rajamani, (2009); Lanyon, Wong, et al., (2015); Walker and Macko (1999), the length of dugong that reaches 2 meters is categorized as an adult over 20 years old. The length and width of the dugong can be seen in Figure 3.

The comparison between the morphometric of dugong (*Dugong-dugon*) in Kabola waters and other morphometric studies in several regions of Indonesia has been adjusted and modified according to Hasan and Tamam, 2019; Iqbal, Setiawan and Yustian, 2020; Hasan and Islam, 2020, as presented in Table 1.

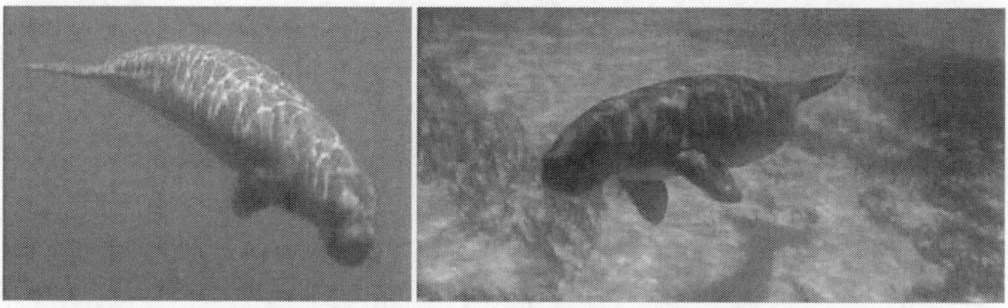

Figure 2. Dugong in seagrass habitat of Kabola Beach Alor.

The behavior of dugong (*Dugong-dugon*) in Kabola waters during the research, from 22 to 24 October 2020, was a very friendly aggressive toward the locals and tourists that were visiting the location. This was shown when dugong popped up quickly to the water surface when it heard the sound of a fleet of motorboats approaching the location. The distance between the boat arrival and the dugong's appearance using Global Positioning System (GPS) tracking was about 40 meters. The illustration of dugong appearance in welcoming the approaching fleet can be seen in Figure 4.

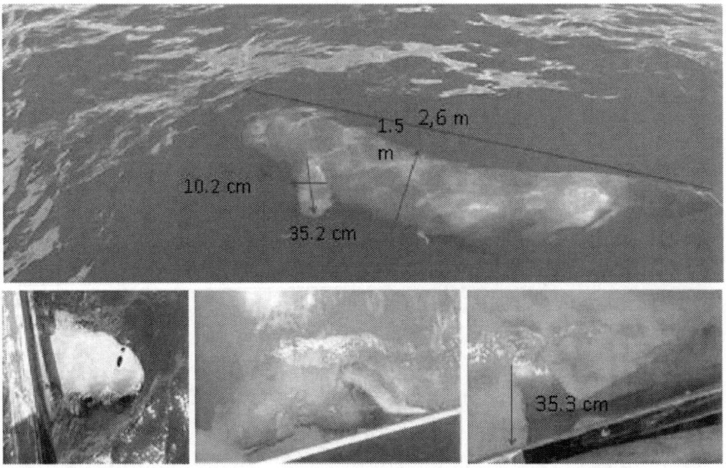

Figure 3. Morphology of dugong in Kabola Beach Alor (Ref: Lanyon et al., 2015).

Table 1. Morphometry of Dugong at Kabola Beach, Alor-Indonesia

Characters	Present Study	MMF-MRMC
1. Length from the Snout to the Base of the Tail	2. 6 m	2.8 m
2. Body Circumference near Navel (*Umbilicus*)	1.5 m	1.5 m
3. Width of the Flipper	10.2 cm	12.3 cm
4. Total Length of the Flipper	35.2 cm	-
5. Width of the Tail	35.3 cm	-

Note: Ministry of Marine Affairs and Fisheries Coastal - Marine Resources Management Centre (MMF-MRMC) Bali, 2018.

The water depth where dugong (*Dugong-dugon*) appears at the research location is about 7 to 10 meters. Dugong (*Dugong-dugon*) appears at the sea surface by raising its nose above the surface for about 3 seconds then entering the water again for approximately 4 to 5 minutes. The same research done by (Nugraha et al., 2019; Tsutsumi et al., 2006; Adulyanukosol et al., 2007) in the waters of Papua-Indonesia, Thailand and Japan concluded that:

Dugong breathes over the sea surface with a time range from 2 to 5 minutes. The appearance of dugong in Kabola-Alor waters can be seen in Figure 4A, B, C and D. The aggressive friendly behavior of dugong (*Dugong-dugon*) in approaching the locals and tourists who visit the location is indicated by the way it approaches the boat and swims back to the water which can be seen in Figure 4 A, B, C and D.

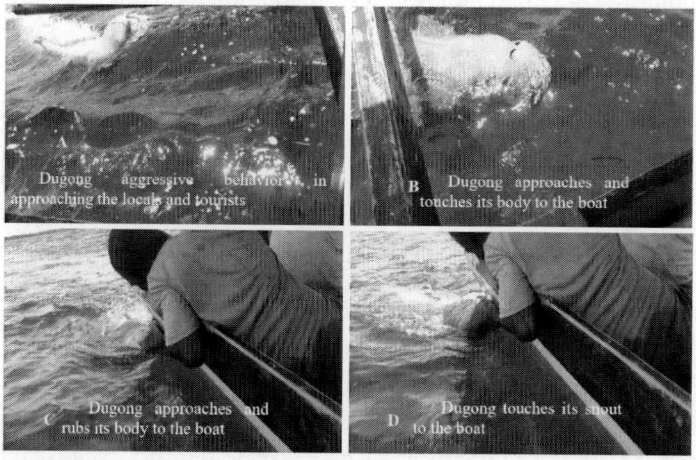

Figure 4. Interaction between dugong and the locals.

Based on the field survey, there are many seagrass beds found in the habitat of dugong at Kabola Beach Alor. Two dominant species of seagrass which are consumed by dugong (*Dugong-dugon*) at Kabola Beach Alor are *Halophila ovalis* and *Enhalus acoroides.* According to Juraij, 2014; Fazillah, 2019; Octavina et al., (2020), the species of seagrass in general consumed by dugong (*Dugong-dugon*) in Indonesian waters are *Halophila ovalis* and *Enhalus acoroides*. The average sea water temperature during this research ranged from 28.3 to 33.70°C.

Whales: Humpback Whale and Stranded Whale

Indonesian waters are inhabited by 31 species of *cetacean* (whale, porpoise, dolphin); twelve of which are whales, while the rest are dolphins and dugongs (*Dugong dugong*) (Rosas et al., 2012). Both are residents and migrants, these fish are distributed throughout the coastal waters to the deep sea zone (Salim, 2011). Several species of cetaceans that are truly migratory use Eastern Indonesian waters as a migration route from the Indian to Pacific Oceans and vice-versa through the waters of the Komodo Islands, Solor-Lembata (East Nusa Tenggara), Banda Sea (*Moluccas*), Southeast Sulawesi, North Sulawesi and Sorong-Fakfak Papua (Salim, 2011). The waters of eastern Indonesia, particularly in several inter-island deep canals, are assumed to serve as the entry point for the migration routes of marine mammals (cetaceans) such as whales and dolphins.

Nowadays, the world's attention is mostly focused on the protection of marine mammals and tends to study the migratory patterns and distribution, especially cetaceans (Dréo et al., 2019). This was due to the decreasing population of cetaceans as a result of the influence of human activities; such as illegal logging, pollution and environmental destruction, thus causing the existence of cetaceans to be protected (Bejder et al., 2019). The humpback whale, *Megaptera novaeangliae*, is a cosmopolitan species that exists in all major ocean basins from tropical to arctic waters (Clapham et al,, 2009; Dawbin, 1959; Secchi et al., 2011). According to IWC 1998; Chittleborough 1965; Findlay, 2011; Secchi et al., 2011, the humpback whale is one of the species that is hunted to be killed and consumed, causing a significant decrease in the species in the southern hemisphere, including the South Atlantic and South America.

The research on whales was conducted in the Pantar Strait, Alor, East Nusa Tenggara from March to December 2020 (Jahved et.al., 2021b). The field survey method was modified by Dharmadi et al., 2017; Savu et al., 2017 to the zig-zag track method using a 10 gross tonnage (GT) boat with two observation decks that was done by two different groups of on-deck observers. Water samples were taken at every appearance of cetaceans to identify the plankton. Field photography documentation of cetaceans appearance was also done for further identification. Field photographs were taken using a Canon camera and drone. The ship was traveling at a speed of 7-8 knots on each survey trajectory. Map of the research area can be seen in Figure 5. On each deck of the ship, monitoring was done using binoculars and each point of appearance of the humpback whale was taken using the global positioning system (GPS) coordinates for further spatial overlay.

Appearances of humpback whales were plotted using GPS and overlaid on data of General Bathymetric Chart of the Oceans data (GEBCO) in the Alor island waters. The data were then processed using the Arcgis 10.8, applying the kriging interpolation method (Jahved et.al., 2021a, 2021b; Wirasatriya et al., 2020; Hartoko, 2010; Hartoko et al., 2019). Data processing was carried out at the Marine Geomatics Laboratory, Diponegoro University, Semarang. The study had discovered three coordinates on the appearance of the humpback whale (*Megaptera novaeangliae*) in the Alor seawaters as presented in Figure 6. While their frequency of appearance is presented in Figure 7.

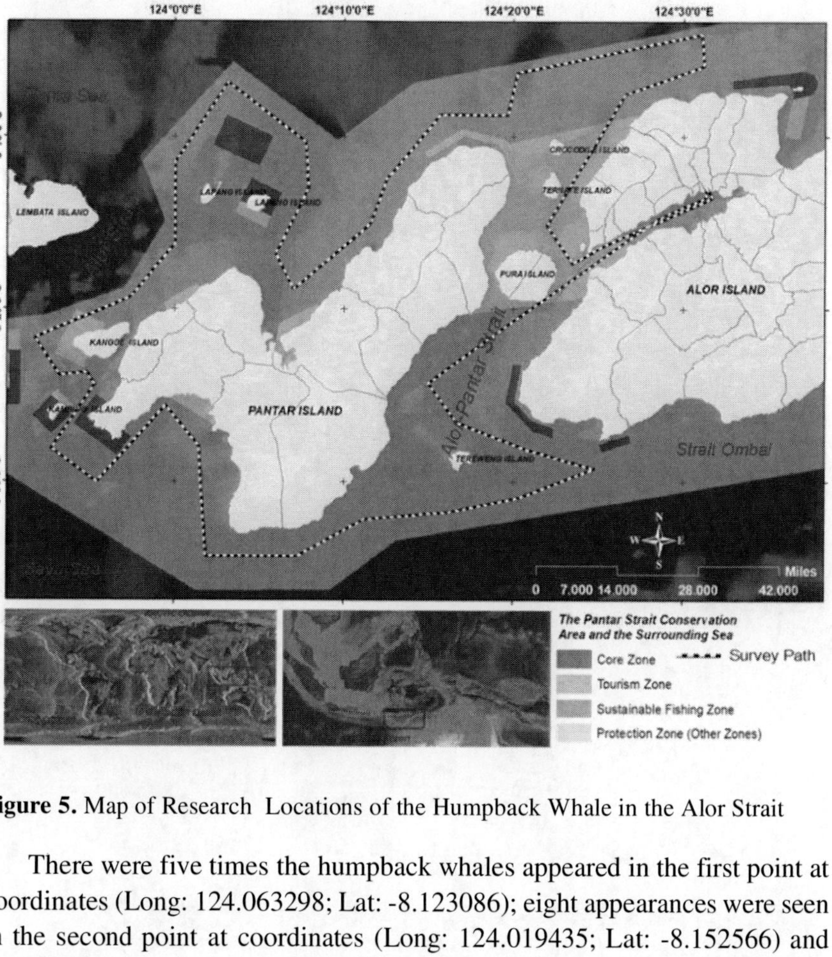

Figure 5. Map of Research Locations of the Humpback Whale in the Alor Strait

There were five times the humpback whales appeared in the first point at coordinates (Long: 124.063298; Lat: -8.123086); eight appearances were seen in the second point at coordinates (Long: 124.019435; Lat: -8.152566) and two appearances were in the third point at coordinates (Long: 124.006817; Lat: -8.249302). The appearance of the humpback whale in Alor waters was found in the afternoon from 05.00-05.45 pm on December 1 2020, which was precisely during the rainy season in Indonesia. Photographs of the humpback whale at each point are presented in Figures 8, 9, and 10.

The appearance of the humpback whale at three points in the Alor Strait waters had the same behavior; jumping over the sea surface, slamming its tail into the water, flipping, and hitting its body in the waterbody which can be seen in Figures 8, 9 and 10.

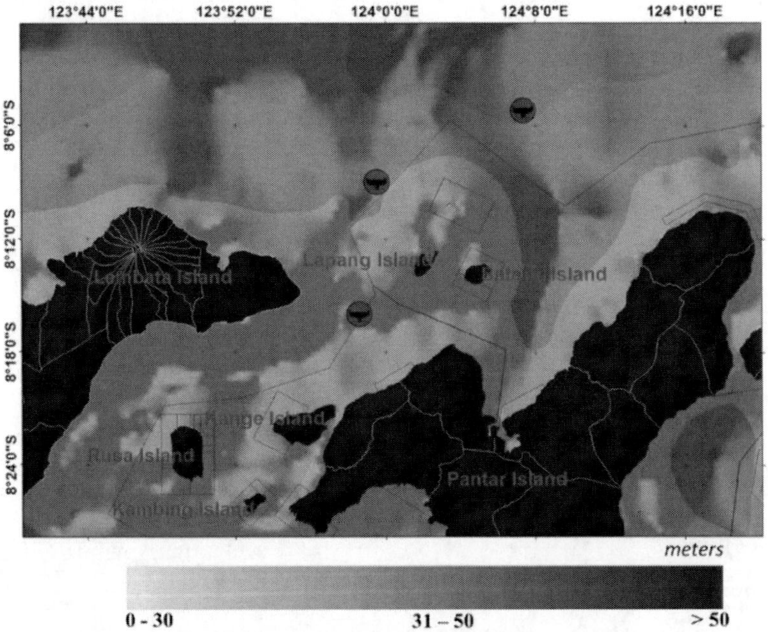

Figure 6. Three coordinates of the appearance of humpback whale (circles).

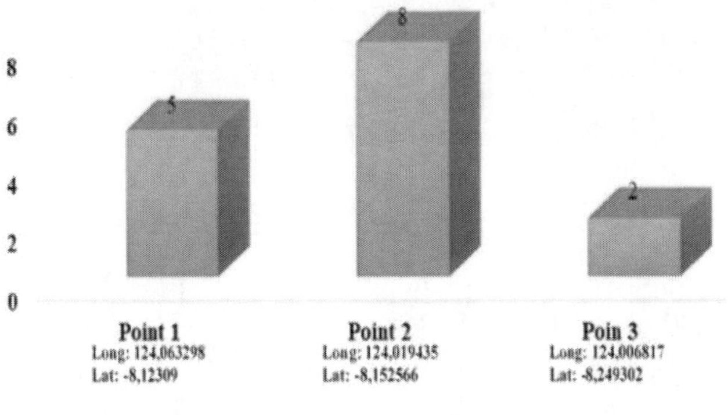

Figure 7. The frequency of appearance of humpback whale (*Megaptera novaeangliae*).

The humpback whale jumped over the sea surface using its elongated pectoral fins during the maneuver and swam back into the water, then came back to the water surface for about five minutes and jumped back for about five seconds and sang (make a sound and appear to the sea surface).

Figure 8. Humpback whale appearance at coordinates of Long: 124.063 E; Lat: 8.123 S.

Figure 9. Humpback whale appearance at coordinates of Long: 124.019 E; Lat: 8.152 S.

Figure 10. Humpback whale appearance at coordinates of Long: 124.019 E; Lat: 8.152 S.

According to Clapham, (2009); Galvin, (2006); Fish & Battle, (1995); Parsons et al., (2008); McSweeney et al., (1989); Zerbini et al., (2004); Calambokidis & Barlow, (2004); Kügler et al., (2020); Mohebbi-Kalkhoran et al. (2019), the humpback whale leaps out of the water (jumping through the seawater), hits the surface of the water with its long fins, then hits the water surface with its tail (lobtail), and freezes in a vertical position with its head above the water (spyhop); it breaks through by swimming from the depth to the surface of the water at an oblique angle, then jumps into the air at an angle (slope of 0-70 degrees) over the sea surface. Then, it rotates on its long axis, landing on its back with its belly rising up, then singing. Singing is an important point of the humpback whale's social behavior with the idea that the sound of its singing has a component of intrasexual selection and / or intersexual selection. The tubercles of the humpback whale in front of its body are used as an enhanced lifting device to control the water flow over its fins and maintain the lift at high angles of attack. The morphology of the humpback whale's fin exhibits high maneuverability related to its unique feeding behavior, whereas its eyeballs are similar to other cetaceans and show adaptation to diving and migration, contributing to the perception of differences in temperature, pressure and lighting (Woodward et al., 2006; Hampe et al., 2015); Rodrigues et al., 2014). The group of the humpback whale that was found in the Alor Strait during the rainy season in December 2020 was assumed to have migrated from the Atlantic Ocean and Australian seawaters. This was due to the appearance of the humpback whale in the Alor Strait waters with a distance of ± 700 kilometers from Australian waters, migrating for food and mating. According to Palsboll, P. J. et al. 1979; IWC.int/humpback-whale (2021, the humpback whales could migrate to find food and mate within a distance of 8,000 kilometers to the breeding grounds in the tropics in the last months of the year which is July-October in the southern hemisphere and December-March in the northern hemisphere. In order to gain enough strength and body mass for giving birth and breastfeeding their babies intensively for several months, the pregnant female humpback whales could swim thousands of kilometers to the nutrient-rich arctic or mid-temperature waters where they could find their food. Usually, the female pregnant humpback whale arrives earlier than other whales which are not pregnant or in adolescence (IWC.int/humpback-whale, (2021). According to Craig et al., (2014) the female pregnant humpback whale seeks for recessed waters (bays) to give birth so that its newborn can be protected. Based on this research, considering the condition of the Alor Strait waters which are quite protected and close to coastal areas. In the case of Alor and its adjacent seawater assumed that one of the reasons of the humpback whale appearance in the Alor Strait waters is due to the rainy season of December 2020.

According to Werth et al., 2019; Burkard et al., (2015); Chen, (2012),the humpback whale has a cranial elevation, which refers to the excretion of filtered water. Cranial elevation of humpbacks begins with a small splash in the mouth anterior, followed by a continuous outflow in the middle or posterior area of the mouth, then releasing the turbulence-free droplet inside the mouth during swallowing. The submersion of the humpback whale's head creates a vortex in the undersea and water surface large enough for the purpose of gathering prey such as groups of small shrimp and small fish to eat. Based on the abundance of zooplankton of the class crustacea at the three points of the humpback whales' appearance, it can be assumed that one of the reasons for its appearance in Alor waters is to migrate for food. This is because the type of plankton from the crustacean class is part of the cetacean diet, including the humpback whale. Oceanographic variables related to the appearance of the humpback whale (Megaptera novaeangliae) in the Alor Strait. Oceanographic variables such as depth, temperature, chlorophyll-a, current velocity are important indicators in supporting the presence of cetaceans in a waterbody (Ballance et al., 2006; Tynan et al., 2005; Cañadas et al., 2002; Hamazaki, 2002; Praca et al., 2009). The appearance of the humpback whale in the Alor Strait waters had an appearance point approximately 7-12 kilometers from the coast, with the depth more than 50 meters. According to Putra et al., (2016); Salim, 2011, cetaceans, especially whales, were found in the Savu Sea waters at a depth less than 100 m. The water depth model is in accordance with the humpback whale found in the Alor Strait waters using GEBCO bathymetry data.

Table 2. Density of zooplankton at three appearance stations/locations of humpback whale at Alor strait

Zooplankton	CLASS	ST 1 (Ind/l)	ST2 (Ind/l)	ST3 (Ind/l)	Total (Ind/l)
Protozoa	Sarcodina	130	127	145	402
Annelida (annulata)	Polychaeta	95	30	72	197
Arthropoda	Crustacea	677	647	633	1957
Mollusca	Gastropoda	112	145	105	362
Echinodermata	Ophiuroidea	6	10	23	39

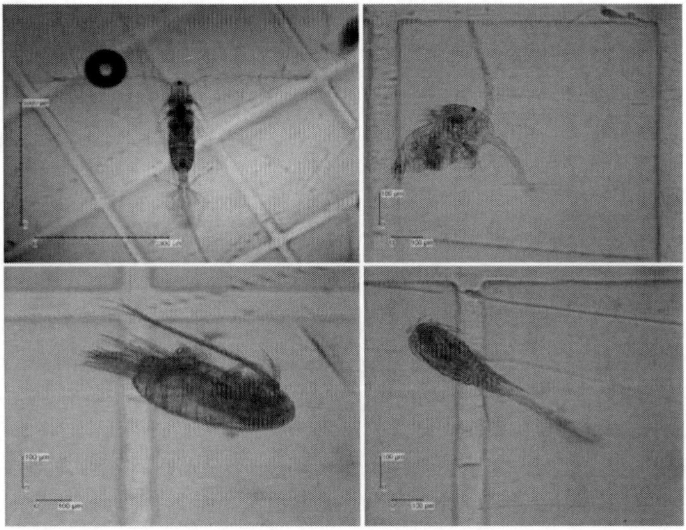

Figure 11. Dominant crustacea found at coordinate point-3 of the humpback whale appearance.

The whale migration into the Savu and Alor Sea was assumed for feeding and reproduction or giving birth. Furthermore, seeing its group behavior and the attraction of throwing its body, then opening its mouth to the surface of the water. Then releasing large bubbles in the waters and drowning itself by poking its tail in the water, it can be assumed that the humpback whale migrated to find food and a mate in the Alor Strait waters. Results of this research found three classes of zooplankton from different phyla at three points of the sighting locations of the humpback whale, along with the identification of zooplankton in the Alor Strait waters. The highest density of zooplankton per-liter was found from the class of crustacea of the phylum Arthropod, with a total of 1957 individuals per-liter found at three observation points, as seen in Table 1 and Figure 11.

Stranded Whales

Almost every year a stranded whale was found at Alor Islands and its adjacent seawaters. One of the worst moments of a group of stranded whale happened in November 7, 2019 at Alor beach. A group of 14 whales of *Physeter macrocephalus* as in Figure 12. Another event of a stranded Blue whale (*Balaenoptera musculus*) happened in July 22, 2020 at Alor beach (Figure 13).

Figure 12. A group of whales *Physeter macrocephalus* stranded at Alor beach in November 7, 2019.

Figure 13. Blue whale (*Balaenoptera musculus*) stranded at Sabu Raijua beach, July 22, 2020.

The phenomena of stranded whales, was found correlated to the phenomena of "cold-seawater" which had been known from local communities, where this had happened since a long time ago. This cold-seawater phenomena also had been scientifically proved by this study using a field sea surface temperature measurement, and also an analysis using a combination of AquaMODIS and Sentinel-3 satellite data (Hartoko,et al., In-press. 2022). There was a sudden drop of sea surface temperature from a normal 30°C to 7-12°C in about 15 minutes during the up welling period.

Figure 14. Whale stranded in December 6.2021 (Courtesy of KCD Alor: Branch Office of Marine and Fisheries, East Nusa Tenggara Province at Alor).

DNA of the Traditional Caught Whale, Stranded Whale and Stomach Content Analysis

Information on the spatial distribution and ecology of whales is very limited, especially for molecular biology or DNA analysis for species identification in Indonesia, meaning that their species distribution and biological behavior is unknown. The research focused on two groups of whales-stranded and caught. This study had revealed that results of DNA analysis confirmed that the stranded whale is *Mesoplodon densirostris*, while the DNA sample of the caught whale is confirmed it as *Physeter macrocephalus*.

According to information of local people around the coast of Alor and Lamalera, *Mesoplodon* (beaked whale) and *P. macrocephalus* are often stranded around the coast of Alor and traditionally become the main source of livelihood food for the local community and whale hunters in Lamalera Island. *P. macrocephalus* with a box-like head morphology is most common in temperate and tropical latitudes and most likely to be found in waters inhabited by squids as they are carnivores for squid or molluscivores. They swim in depths from 1,000 m to the surface of the sea and likely with cold-water upwellings. Because they are so well-adapted for deep water swimming, they are in real danger of stranding when they move inshore. The stomach contents analysis of both stranded and caught whales were based on the samples of the digested material found in their stomachs. DNA analysis revealed that squid of *Sepioteuthis lessoniana* is the whales' natural diet found in their stomach. Tracing the natural diet of whales in the Strait of Alor and Lamalera using DNA analysis of the whale stomach contents indicated that their food in the area predominantly consists of squid from the species *S. lessoniana* (Pringgenies, et al., 2021). A study by Clarke and Pascoe (1997), on the analysis of stomach contents of stranded, male *P. catodori* whales in Penzance, Cornwall reported a diet consisting of 4 families of cephalopods, namely *Histioteuthis* spp., *Architeuthis* spp., *Todarodes sagittatus* and *Galiteuthis sp.* This diet is enriched with the intake of Oceanic Octopod, *Haliphron atlanticus* and only one neritic species of *Loligo forbesi*. Results of this study indicated that most of the squid consumed were oceanic species, namely *Histioteuthis* spp., *Architeuthis* spp., *Todarodes sagittatus* and *Galiteuthis sp.* This finding is similar to what was reported in a study by Santos et al., (2001) that discovered that the whale's diet consisted of oceanic squid, oceanic fish remains and crustaceans. The squid *S. lessoniana* found in the stomach of *P. macrocephalus* from the coast of Alor Indonesia, is known as a neritic species. This finding could be an indication that the coasts of Alor and Lamalera are the feeding grounds for these two whale species because of the concentration of the squid species number and the suitability of the currents in the two regions.

Whale migration and residence patterns are known to be closely related to currents, which in turn affect the potential abundance of food sources in the seawaters (Arkhipkin et al., 2015). This condition makes the straits of Alor and Lamalera the ideal stopping points in the migration routes of these two whale species. *S. lessoniana* is a squid species with a very short life cycle, only

4 months, with more than 1000 spawns being hatched per cycle. After they lay eggs, they still have a chance to spawn again. This fertility is the source of abundance and the speed of population recovery for this species (Pringgenies, et al., 2000, Arkhipkin et al., 2015).

In comparison, the squid of the *Loligo duvaucelii* species has a life cycle of 12 months. Mature individuals of this species only have the opportunity to spawn eggs once, because afterward, the parent will die (Hanlon, 1998, Lipiński, Vyver, Shaw, and Sauer, 2016). Visual and field observations at the coastal fish markets of Alor and Lamalera showed abundant populations of squid as well as fish species from deep waters in these two areas. Jefferson, Leatherwood, Webber (1993) estimated that *M. densirostris* has various species of deep-sea fish and squids as its natural diet. Furthermore, it was also reported in the paper that Blainville's beaked whale is found in waters of temperate and the tropical area. The species is commonly found off the shores of Hawaii, where they have been observed both in the water and stranded on the coast. They are found in deep water, either individually or in small pods, where the depth of the oceans reaches 1,000 fathoms (about 660 m).

Due to their deep-water habitat, over the years Blainville's beaked whale has rarely been seen swimming freely, only identified as a species by its skeletal remains. The identification results by molecular methods of caught whales indicated that the species was *Physeter macrocephalus* since limited information is available on this species. The baleen whale has been seen in the waters around Denmark with a length of almost 9 meters, possessing a very large head covering almost its entire body (Friel and Sleeman, 2003) similar to those seen in the waters of Alor and Lamalera, Indonesia. An increase in *P. macrocephalus* population was reported among waters in England and Ireland (Cabot, 1967). The information is very interesting because this could be meaning that the whales of the species *P. macrocephalus* may thrive there in spite of their low rate of mortality. According to Nybakken (1992) most of the zooplankton is class crustacean and belongs to the phylum of Arthropod, the only crustaceans that live as plankton. They are the most important zooplankton for fish in freshwater and sea water. Crustaceans are animals that have cells consisting of chitin or calcium which are difficult to digest. Crustaceans can be divided into 2 groups: Entomostraca or low-level crustaceans and Malacostracea or high-level crustaceans.

Alor Strait, which is still influenced by the water mass of the Indian Ocean and the Pacific Ocean, makes Alor waters quite sensitive to changes in sea water temperature, especially at the point where the humpback whale appears in Alor waters. Research results through ARGO (Array for Real-Time

Geostrophic Oceanography) data analysis showed that the monthly average of sea water temperature from 2006-2020 at a depth of 0-200 meters ranged from 18.0-28.0°C which can be seen in Figure 8. Meanwhile, the monthly average of sea surface temperature data processed through Aqua MODIS (Moderate Resolution Imaging Spectroradiometer) data ranged from 25.0-31.0°C. Using ARGO and Aqua MODIS data, the research found that the average depth and sea surface temperature in the Alor Strait waters including its surrounding seas in the tropics are still in the normal category for the emergence of cetaceans, especially the humpback whales. This is supported by the research of Houser et al., (2004) which explains that the average temperature range in the tropics for the appearance of cetaceans is between 20-31°C. Forcada et al., (1996) documented the preferred temperature for cetaceans to be between 22.3-26.3°C (average 24.2°C).

The same case was conveyed by (Laran and Gannier, 2008; Gregr dan Trites, 2001; Hamazaki, 2002; Doniol-Valcroze et al., 2007) that consistently, sea surface temperature and depth for the presence of cetaceans in water bodies ranges from 22.4-26.7°C (average 23.08°C). This is also supported by Putra et al., (2016) found the appearance of cetaceans in the Savu Sea waters in the range of 30-31°C sea urface temperature. The average distribution of chlorophyll-a and the average monthly sea surface flow velocity based on the results during 2020 based on Aqua MODIS data analysis in the Alor Strait seawater ranged from 0.2-0.8 $mm.m^3$ with current velocities ranging from 0.2-0.6 $m.s^{-1}$. The research of found that the average range of chlorophyll-a and current velocities for the emergence of cetaceans in Alor waters ranging from 0.1-0.6 mm. m^{-3} for chlorophyll-a, with mean sea surface flow velocities ranging from 0.2-0.6 $m.s^{-1}$. To this extent, it can be concluded that the average range of chlorophyll-a and the average flow velocity in the Alor Strait are still categorized as normal for the schooling and appearance of cetaceans in the waters of the Alor Strait.

The whale appearance location, species, behaviour, appearance frequency and time during the east season June-November 2020 is presented in Table 3. Their appearance had been detected in 8 locations, which are South Alor, East of Batang Island, South Lembata Island, North Lembata Island, Savu Sea, West of Kupang Bay, North of Savu Sea and at Ombay Strait. Three whale species had been identified are *Kogia breviceps, Balaenoptera musculus, and Physeter macrocephalus*. The whale appearance coordinates during the east season is presented in Figure 15.

Table 3. Whale appearance location, species, behaviour and appearance during the east season (June-November 2020)

Location	Species		No.of Appear	Appear Time (hr)
South Alor Island	Kogia breviceps	Lobtailing, Feeding	1	0.02
East of Batang Island	Balaenoptera musculus	Lobtailing, Feeding	2	0.03
South Lembata Island	Physeter macrocephalus	Lobtailing dan Avoidance	1	0.02
North Lembata Island	Physeter macrocephalus	Lobtailing, Feeding	1	0.02
Savu Sea	Balaenoptera musculus	Lobtailing	2	0.03
Savu Sea	Kogia breviceps	Lobtailing, Feeding	1	0.02
West of Kupang Bay	Kogia breviceps	Feeding	1	0.02
North of Sawu Sea	Physeter macrocephalus	Breaching	1	0.02
Ombai Strait	Balaenoptera musculus	Lobtailing	2	0.03

Tabel 4. Whale appearance location, species and behaviour west season December 2020 – April 2021

Locations	Species		No.of Appear	Appear Time (hour)
North East Alor	Balaenoptera musculus	Lobtailing, Feeding	1	0.02
Lapang Island	Megaptera novaeangliae	Avoidance	5	0.08
Alor Flores Strait	Megaptera novaeangliae	Avoidance	10	0.17
South Lembata Island	Physeter macrocephalus	Avoidance	1	0.02
Pantar Strait	Physeter macrocephalus	Lobtailing, Feeding	1	0.02

Five types of appearance behaviour had been detected during the east season, with average period of appearance of about 0.02 hours (1.2 minute). While during the west season (December 2020 – April 2021) whale appearances were detected at five locations- Northeast Alor, Lapang Island, Alor-Flores Strait, South Lembata and Pantar Strait as in Table 4.

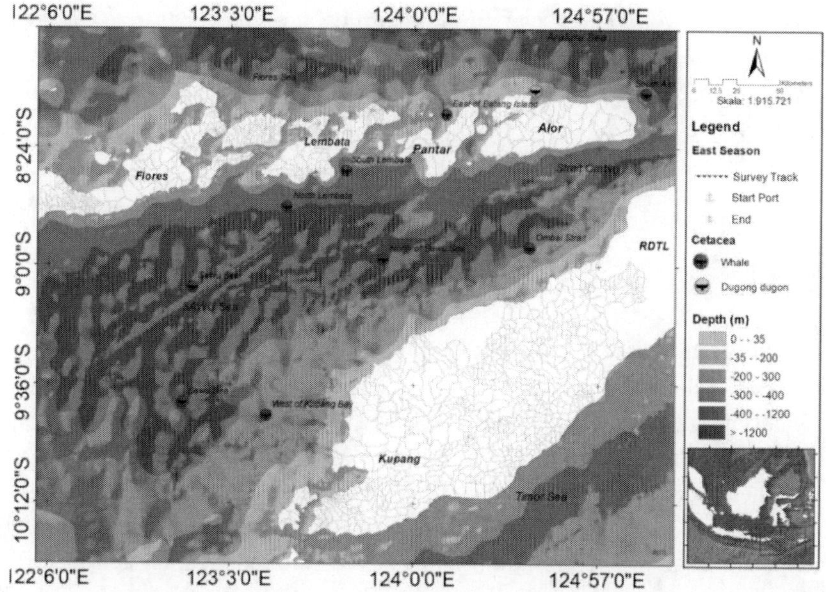

Figure 15. Whale appearance sighting during east season June – November 2020.

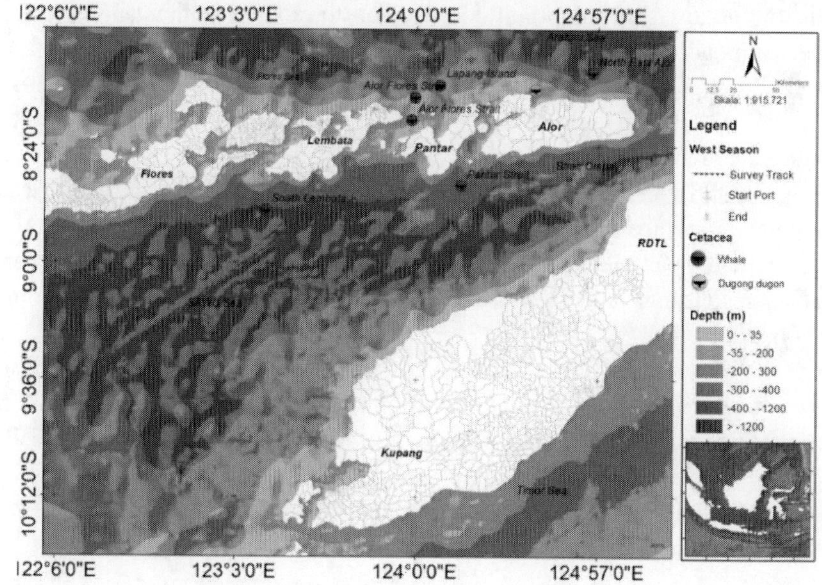

Figure 16. Whale appearance sighting during west season December 2020 – April 2021.

Three species of whales had been identified are *Balaenoptera musculus, Megaptera novaengliae,* and *Physeter macrocephalus.* Two appearance behaviours detected are Lobtailing-feeding and avoidance. The average time of appearance is 0.02 hours (1.2 minute). The whale appearance coordinates during the east season is presented in Figure 16.

Dolphin

Introduction

Indonesian waters are inhabited by 31 types of cetaceans, (whale, porpoise, and dolphin), twelve among them are whales and the rest are dolphins, and others are dugongs (Rosas et al., 2012; WWF, 2020; Yusmalinda et al., 2017). Few types of cetaceans that are migrants or nomads use the Eastern of Indonesian waters as a migration route between the Indian and Pacific Oceans through the waters of the Komodo Island, Solor-Lembata (East Nusa Tenggara), Banda Sea (Maluku), Southeast Sulawesi, North Sulawesi and Sorong-Fakfak (Papua) (Dréo et al., 2019; Goetz et al., 2015). The record of aquatic biotais is an important contribution to raise an understanding of the species biogeography and diversity in Indonesia (Hasan et al., 2019a; Hasan et al., 2019; Valen et al., 2019; Saptadjaja et al., 2019; Widodo et al., 2020; Hasan et al., 2021; Hasan et al., 2021). One seawater area in Indonesia where cetaceans are often found is the Pantar Strait in Alor Regency, East Nusa Tenggara Province. These waters are unique in that there are significant changes in surface temperature and salinity during the southeast monsoons. Water quality parameters have an important role in the conservation of aquatic organisms (Islamy and Hasan, 2020; Hasan and Widodo, 2020; Hasan and Islam, 2020).

The dynamics of these seawater masses is that they occur at the surface layer which is influenced by monsoon wind patterns. These conditions resulted in upwelling in the waters of the Pantar Strait. The process of the seawater mixing and upwelling will affect the living conditions of phytoplankton, hydrology and nutrient enrichment in these waters (Sydeman et al., 2014; García-Reyes et al., 2015).

Method

The research was conducted in Alor- Pantar Strait, Alor Regency, East Nusa Tenggara Province on May 28-29 2020. The field survey method was modified by Dharmadi et al., (2017); Savu et al., (2017) namely by a zig-zag route using ships with 10 GT (gross tonnage) with 2 observation decks and 2 observer groups on each deck. Every appearance of dolphins is documented in the form of an image so that it can be identified. Pictures were taken using Canon cameras and drones. The ship was traveling at a speed of 7-8 knots on each survey track. On each deck of the ship, monitoring is carried out using binoculars and each point of appearance of the cetacean is noted with the coordinates using global positioning system (GPS) for data analysis.

The oceanographic data processing uses Aqua Modis satellite imagery data, Copernicus Marine Service Data Argo Flot, GEBCO data (General Bathymetric Chart of the Oceans) in the waters of Alor Strait. The data were processed using Arcgis 10.8 using the kriging interpolation method (Wirasatriya et al., 2020; Hartoko et al., 2008). Types of data processed in this research are sea surface temperature data, chlorophyll-a, current speed data, wind speed data and depth data. Data processing was carried out at the Marine Geomatic Laboratory Faculty of Fisheries and Marine Science of Diponegoro University, Semarang.

Results

The field survey found 9 points of dolphin appearances with the total length of the survey path were the same as the GPS measurement of 99.72 kilometers. Two species of dolphins found are the bottlenose dolphin (*Tursiops truncates*) and the spinner dolphin (*Stenella longirostris*) as presented in Figure 17. The sighting behaviors are bow riding, aerials, stationary, traveling, lobtailing and avoidance. The locations, species, behaviours and appearance time during the east season as in Table 5 and for the west season is presented in Table 6. The position of the dolphin appearance during the east season is presented in Figure 18 and for the west season as presented in Figure 19. The group appearance of the bottlenose dolphin ranges from 3-50 individuals. The body shape of the dolphin is torpedo like or streamlined and it is very hydrodynamic with a tapered and slender tip so that it can easily move without significant obstacles in the water (Leeney et al., 2007; Rako et al., 2013).

Figure 17. A & B. The appearance of bottlenose dolphin (*Tursiops truncates*) in the waters of the Pantar Strait; C & D The appearance of spinner dolphin (*Stenella longirostris*) and *Tursiops truncates* in the waters of the Pantar Strait; E & F Sightings of dolphins at an altitude of 15 meters using drones at Pantar Strait. (Documentation: Jahved F Maro and Mike, May-2020).

The depth range of the appearance of bottlenose dolphins and spinner dolphins ranges from 20-50 m overlaid using GEBCO (General Bathymetric Chart of the Oceans) bathymetry data. Monthly mean sea surface temperature based on AquaMODIS data in Alor-Pantar Strait in 2020 is in the range 28.0-31.0^{0}C. Meanwhile, the monthly average range of chlorophyll-a using AquaMODIS data of 2020 ranges is 0.2-0.8 mm/m^3, average sea surface current velocity in 2020 ranging from 0.2-0.9 m.s^{-1}.

Table 5. Location, species, behaviour and appearance of dolphins during the east season June-November 2020

Location	Species	Behaviour	No.of Appear	Appear Time (hr)
Alor Kecil	Bottlenose Dolphin (Tursiops truncatus)	Lobtailing, Avoidance	2	0.25
East Pura	Bottlenose Dolphin (Tursiops truncatus)	Lobtailing, Avoidance	2	0.25
Batu Putih	Bottlenose Dolphin (Tursiops truncatus)	Lobtailing, Avoidance	2	0.28
Pantar Strait	Bottlenose Dolphin (Tursiops truncatus)	Lobtailing, Avoidance	2	0.28
Pantar Strait	Bottlenose Dolphin (Tursiops truncatus)	Lobtailing, Feeding	2	0.42
Lapang Island	Bottlenose Dolphin (Tursiops truncatus)	Lobtailing, Feeding	2	0.30
Puntaru	Bottlenose Dolphin (Tursiops truncatus)	Lobtailing, Feeding	2	0.42
Marica	Bottlenose Dolphin (Tursiops truncatus)	Lobtailing, Feeding	2	0.18
Batang Island	Bottlenose Dolphin (Tursiops truncatus)	Lobtailing, Feeding	2	0.18
Alor-Flores Strait	Spinner dolphin (Stenella longirostris)	Lobtailing, Feeding	2	0.38
Lembata	Spinner dolphin (Stenella longirostris)	Lobtailing, Avoidance	2	0.42
Savu Sea	Spinner dolphin (Stenella longirostris)	Lobtailing, Avoidance	2	0.28
West Kupang Bay	Spinner dolphin (Stenella longirostris)	Lobtailing, Feeding	2	0.28
West Kupang Bay	Spinner dolphin (Stenella longirostris)	Lobtailing, Feeding	2	0.18
Lamalera	Spinner dolphin (Stenella longirostris)	Lobtailing, Feeding	2	0.25
Savu Sea	Spinner dolphin (Stenella longirostris)	Lobtailing, Feeding	2	0.18
Savu Sea	Risso's dolphin (Grampus griseus)	Lobtailing, Feeding	1	0.22
South Lamamera	Risso's dolphin (Grampus griseus)	Avoidance, Feeding	1	0.28

Table 5. (Continued)

Location	Species	Behaviour	No.of Appear	Appear Time (hr)
Savu Sea	Risso's dolphin (Grampus griseus)	Lobtailing, Avoidance	1	0.25
Munaseli	Bottlenose Dolphin (Tursiops truncatus)	Lobtailing, Avoidance	1	0.25
Savu Sea	False killer whale (Pseudorca crassidens)	Avoidance, Feeding	1	0.18
Ombai strait	False killer whale (Pseudorca crassidens)	Lobtailing, Avoidance	1	0.42
Ombai strait	Spinner dolphin (Stenella longirostris)	Lobtailing, Avoidance	2	0.18
Savu Sea	Spinner dolphin (Stenella longirostris)	Lobtailing, Avoidance	2	0.18

Table 6. Location, species, behaviour and appearance of dolphin during the dolphin west season December 2020-May 2021

Location	Species	Behaviour	No.of Appear	Time Appear (hr)
Alor Kecil	Tursiops truncatus	Lobtailing, Feeding	2	0.28
Alor	Tursiops truncatus	Lobtailing, Feeding	2	0.18
Pantar Strait	Tursiops truncatus	Lobtailing, Feeding	2	0.25
East Tereweng	Stenella longirostris	Avoidance	2	0.25
Ombai Strait	Stenella longirostris	Lobtailing	1	0.22
South Pantar	Stenella longirostris	Avoidance	1	0.18
South Lembata	Tursiops truncatus	Lobtailing, Avoidance	2	0.23
East Lembata	Tursiops truncatus	Lobtailing, Feeding	2	0.30
South Rusa	Tursiops truncatus	Lobtailing, Feeding	3	0.32
West Rusa	Tursiops truncatus	Lobtailing, Feeding	3	0.33
East Lembata	Tursiops truncatus	Avoidance	1	0.25
North Lapang	Stenella longirostris	Lobtailing, Avoidance	1	0.33
North Batang	Stenella longirostris	Avoidance	1	0.15
North Munaseli	Tursiops truncatus	Avoidance	1	0.27
North Munaseli	Tursiops truncatus	Lobtailing	1	0.33

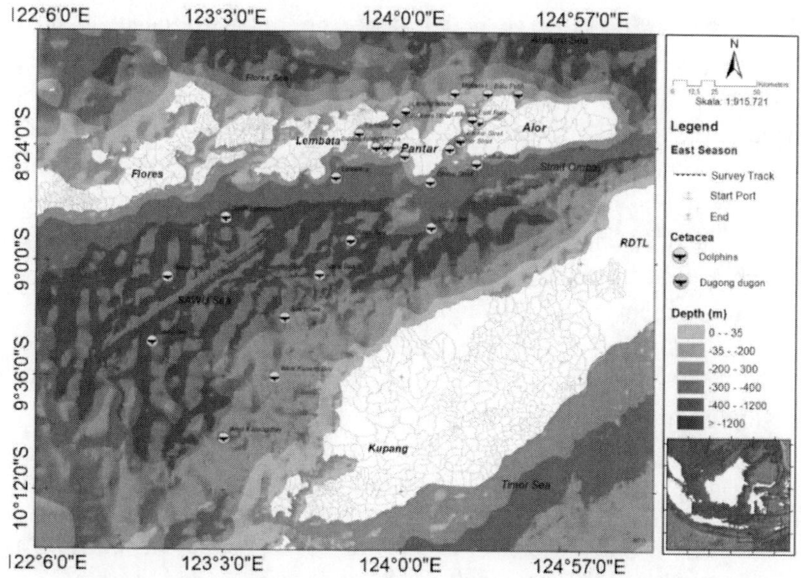

Figure 18. Dolphin appearances during the east season June-November 2020.

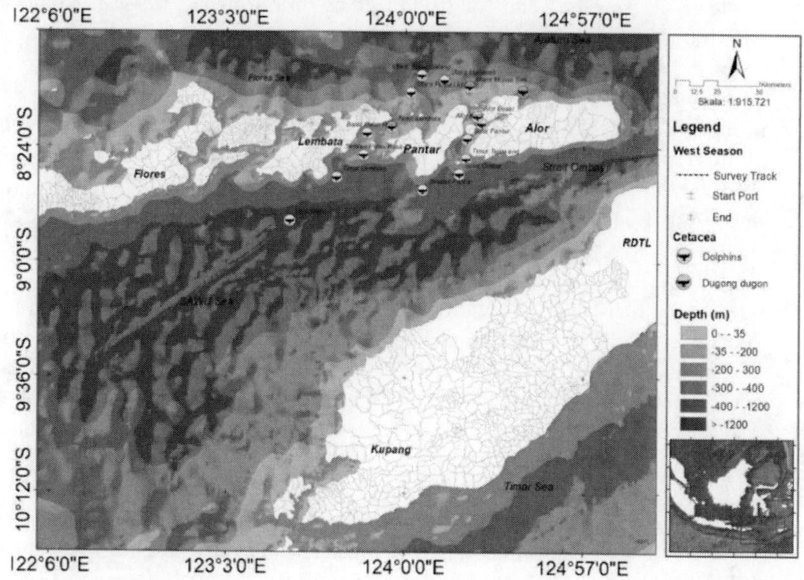

Figure 19. Dolphin sightings during the west season December 2020-May 2021.

In general, the average range of oceanographic variables in the Alor-Pantar Strait is considered in a normal range in the tropics for the appearance of dolphins. Houser et al., (2004) explained that the average of temperature range in the tropics for the appearance of dolphins is between 20-31^{0}C. The study of Putra et al., (2016) in Savu seawaters, found bottlenose dolphins and spinner dolphins in the temperature range of 30-31°C with the depth less than 100m. The difference of the seawater temperature is considered to affect the appearance of cetaceans. This was since some cetacean can maintain their metabolism in low and high temperature ranges.

References

Adulyanukosol. Kanjana, Surasak Thongsukdee, Takeshi Hara, Nobuaki Arai, Makoto Tsuchiya. 2007. Observations of dugong reproductive behavior in Trang Province, Thailand: Further evidence of intraspecific variation in dugong behavior,' *Marine Biology*. doi: 10.1007/s00227-007-0619-y.

Aketa, K. Shiro Asano, Yoshihito Wakai, Akito Kawamura. 2003. Apparent digestibility of eelgrass Zostera marina by captive dugongs. Dugong in relation to the nutritional content of eelgrass and dugong feeding parameters. *Mammal Study*. doi: 10.3106/mammalstudy.28.23.

Anderson, P. 1981. The behavior of the dugong (*Dugong dugon*) in relation to conservation and management,' *Bulletin of Marine Science*.

Aragones, L. V. Ivan R Lawler, William J Foley, Helene Marsh. 2006. Dugong grazing and turtle cropping: Grazing optimization in tropical seagrass systems?', *Oecologia*. doi: 10.1007/s00442-006-0477-1.

Arkhipkin A. I., Rodhouse P. G. K., Pierce G. J., Sauer W., Sakai M., Allcock L., Arguelles J., Bower J. R., Castillo G., Ceriola L., 2015 World squid fisheries. *Rev Fish Sci Aquac* 23(2):92–252.

Blair, D. Adrian McMahon, Brenda McDonald, Daniela Tikel, Michelle Waycott, Helene Marsh. 2014. 'Pleistocene sea level fluctuations and the phylogeography of the dugong in Australian waters,' *Marine Mammal Science*. doi: 10.1111/mms.12022.

Balance, L. T., Pitman, R. L. and Fiedler, P. C. 2006. Oceanographic influences on seabirds and cetaceans of the eastern tropical Pacific: A review. *Progess in Oceanography* 69: 360-390.

Bejder, L., Vidensen, S., Hermensen, L., Simon, M., Hanf, D. and Medsen, P. T. 2019. Low energy expenditure and resting behaviour of humpback whale mother-calf pairs highlights conservation importance of sheltered breeding areas. *Scientific Report*. 9(711): 1-11.

Burkard, M., Whitworth, D., Schirmer, K. and Nash, S. B. 2015. Establishment of the first humpback whale fibroblast cell lines and their application in chemical risk assessment. *Aquatic Toxicology* 167: 240-247.

Cabot D., 1967 A Sperm Whale, Physeter catodon (L.) on Achill Island. *Irish Nat J* 15(11):15–16.

Calambokidis, J. and Barlow, J. 2004. Abundance of blue and humpback whales in the eastern north pacific estimated by capture-recapture and line-transect methods. *Marine Mammal Science*. 20(1): 63-85.

Cañadas, A., Sagarminaga, R. and García-Tiscar, S. 2002. Cetacean distribution related with depth and slope in the Mediterranean waters off southern Spain. *Deep Sea Research Part I: Oceanographic Research Papers*. 49(11): 2053-2073.

Chen, J.-H. 2012. The effect of humpback whale-like leading edge protuberances on hydrofoil performance. *15th International Symposium on Flow Visualization*. 2012: 1-76.

Chittleborough RG. 1965. Dynamics of two populations of the humpback whale, Megaptera novaeangliae (Borowski). *Australian Journal of Marine and Freshwater Research* 16:33–128.

Clapham, P. J. 2009. Humpback Whale: Megaptera novaeangliae. *Encyclopedia of Marine Mammals* (Second Edition). Academic Press: 582-585.

Craig, A. S., Herman, L. M., Pack, A. A. and Waterman, J. O. 2014. Habitat segregation by female humpback whales in Hawaiian waters: avoidance of males? *Behavior*. 151(5): 613-631.

Clarke M. R., Pascoe P. L., 1997 Cephalopod species in the diet of a sperm whale (Physeter catodon) stranded at penzance, Cornwall. *J Mar Biol Assoc* 77:1255–1258.

Dewi, C. S. U, Sukandar, Beginer Subhan, Dondy Arafat. 2018. Distribution of the Dugong Feeding Habitat and Its Threat in Small Islands of INDONESIA,' *JFMR-Journal of Fisheries and Marine Research*. doi: 10.21776/ub.jfmr.2018.002.02.9.

Dharmadi, D., Faizah, R. and Wiadnyana, N. N. 2017. Frekuensi pemunculan, tingkah laku, dan distribusi mamalia laut di laut Sawu, Nusa Tenggara Timur [Appearance frequency, behavior, and distribution of marine mammals in the Savu Sea, East Nusa Tenggara]. *Bawal Widya Riset Perikanan Tangkap*. 3 : 209-216.

Dréo, R., Bouffaut, L., Leroy, E., Barruol, G. and Samaran, F. 2019. Baleen whale distribution and seasonal oc- currence revealed by an ocean bottom seismometer network in the Western Indian Ocean. *Deep Sea Research Part II: Topical Studies in Oceanography* 161: 132-144.

Evans, P. G. H.. 2008. Proceedings of the Workshop On Offshore Wind Farms and Marine Mammals: Impacts and Methodologies for Assessing Impacts', *European Cetacean Society 21st Annual Conference. European Cetacean Society, San Sebastian, Spain*.

Fazillah M. R. 2019. *Biodiversity of Seagrass as Potential Feeding for Dugong at Lamteng Bay Aceh*. ETD Unsyiah.

Fish, F. E. and Battle J. M. 1995. Hydrodynamic design of the humpback whale flipper. *Journal of Morphology*. 255(1): 51-60.

Findlay K., Meÿer M., Elwen S., Kotze D., Johnson R., Truter P., Uamusse C., Sitoe S., Wilke C., Kerwath S., Swanson S., Staverees L., Westhuizen J. V. D., 2011 Distribution and abundance of humpback whales, Megaptera novaeangliae, off the coast of Mozambique. *Journal of Cetacean Research and Management* 3:163–174.

Friel D., Sleeman D. P., 2003 Cetacean Notes Sperm whale Physeter catodon (Meyen) Minke whale Balaenoptera acutorostrata Lacepede. *Irish Nat J* 27(8):322.

Galvin, C. 2006. Surface-Piercing Activities of the Humpback Whale, Megaptera, Related to Parasites and Mechanics. *American Geophysical Union*. 87(52).

García-Reyes, M., Sydeman, W. J., Schoeman, D. S., Rykaczewski, R. R., Black, B. A., Smit, A. J. and Bograd, S. J. 2015. Under pressure: Climate change, upwelling, and eastern boundary upwelling ecosystems. *Frontiers in Marine* Science 2: 109.

Garrigue, C., Patenaude, N. and Marsh, H. 2008. 'Distribution and abundance of the dugong in New Caledonia, southwest Pacific,' *Marine Mammal Science*. doi: 10.1111/j.1748-7692.2007.00173.x.

Goetz, S., F. L. Read, M. Ferreira, J. M. Portela, M. B. Santos, J. Vingada, U. Siebert, A. Marcalo, J. Santos, S. Monteiro, M. Caldas, M. Riera and G. J. Pierce. (2015) 'Cetacean occurrence, habitat preferences and potential for cetacean-fishery interactions in Iberian Atlantic waters: Results from cooperative research involving local stakeholders,' *Aquatic Conservation: Marine and Freshwater Ecosystems*. doi: 10.1002/aqc.2481.

Grech, A., Sheppard, J. and Marsh, H. 2011. 'Informing species conservation at multiple scales using data collected for marine mammal stock assessments,' *PLoS ONE*. doi: 10.1371/journal.pone.0017993.

Grech A, Sheppard J, Marsh H. Informing species conservation at multiple scales using data collected for marine mammal stock assessments. *PLoS One*. 2011;6(3).

Gregr E. J., Trites A. W., 2001 Predictions of critical habitat for five whale species in the waters of coastal British Columbia. *Canadian Journal of Fisheries and Aquatic Sciences* 58:1265-1285.

Hamazaki, T. 2002. Spatiotemporal prediction models of cetacean habitats in the mid-western North Atlantic Ocean (from Cape Hatteras, North Carolina, U.S.A. to Nova Scotia, Canada). *Marine Mammal Science*. 18(4): 920-939.

Hampe, O., Franke, H., Hipsley, C. A., Kardjilov. and Müller, J. 2015. Prenatal cranial ossification of the humpback whale (Megaptera novaeangliae). *Journal of Morphology*. 276(5): 564:82.

Hartoko, A., Wirasatria, A., Helmi, M. and Rochaddi, B. 2008. Aplikasi Teknologi Geomatika Untuk Pemetaan Penurunan Tanah (Land-Subsidence) di Pesisir Kota Semarang Aplication of Geomatic Tech- nology for Land-subsidence Mapping of Semarang Coastal City. *Indonesian Journal of Marine Science*. 13 (1): 13-18.

Hartoko, A. 2010. Spatial distribution of Thunnus.sp, vertical and horizontal sub-surface multilayer temperature profiles of in-situ agro float data in Indian Ocean. *Journal of Coastal Development*. 14(1): 61-74.

Hartoko, A., A. Febrianto, A. Pamungkas and I. Fachruddin. (2019) 'The myth and legend of Sadai and Gaspar strait Bangka Belitung (Banca-Billiton) and oceanographic conditions,' *International Journal of GEOMATE*, 17(62), pp. 212–218. doi: 10.21660/2019.62.93965.

Hasan, V., Soemarno., Widodo, M. S. and Wiadnya, D. G. R. 2019a. *Lobocheilos falcifer* (Valenciennes, 1842) (Cyrpiniformes, Cyprinidae): distribution extension in Java and first record from Tuntang River, Semarang Regency, Indonesia. *Ecology, Environment and Conservation*. 25 (4) : 1713-1715.

Hasan, V., Soemarno, Widodo, M. S., Wiadnya, D. G. R., Mukti AT. and Irawan B. 2019b. Distribution extension and first record of *Lobocheilos falcifer* (Cypriniformes,

Cyprinidae) in Central Java Province, Indonesia. *Ecology, Environment and Conservation.* 25 : 158-161.

Hasan, V. and Widodo, M. S. 2020. Short Communication: The presence of bull shark *Carcharhinus leucas* (Elasmobranchii: Carcharhinidae) in the fresh waters of Sumatra, Indonesia. *Biodiversitas.* 21 : 4433- 4439.

Hasan, V. and Islam, I. 2020. First inland record of bull shark *Carcharhinus leucas* (Müller & Henle, 1839) (Carcharhiniformes: Carcharhinidae) in Celebes, Indonesia. *Ecologica Montenegrina.* 38 : 12-17.

Hasan, V., Gausmann, P., Nafisyah, A. L., Isroni, W., Widodo, M. S., Islam, I. and Chaidir, R. R. A. 2021a. First record of Longnose marbled whipray *Fluvitrygonoxyrhyncha* (Sauvage, 1878) (Myliobatiformes: Dasyatidae) in Malaysian waters. *Ecologica Montenegrina.* 40 : 75–79.

Hasan, V., Valen, F. S., Islamy, R. A., Widodo, M. S., Saptadjaja, A. M. and Islam, I. 2021b. Short Communication: Presence of the vulnerable freshwater goby *Sicyopusauxilimentus* (Gobiidae, Sicydiinae) on Sangihe Island, Indonesia. *Biodiversitas.* 22 : 573–581.

Hasan, V. and Islam, I. 2020. First inland record of bull shark Carcharhinus leucas (Müller & Henle, 1839) (Carcharhiniformes: Carcharhinidae) in Celebes, Indonesia. *Ecologica Montenegrina*. doi: 10.37828/em.2020.38.3.

Hasan, V. and Tamam, M. B. 2019. 'First record of the invasive Nile Tilapia, Oreochromis niloticus (Linnaeus, 1758) (Perciformes, Cichlidae), on Bawean Island, Indonesia', *Check List*. doi: 10.15560/15.1.225.

Hodgson, A. J. 2009. 'Chapter 8 : Marine Mammals' *Marine Atlas of the Western Arabian Gulf.*

Houser, D. S., Finneran, J., Carder, D., Van Bonn, W., Smith, C., Hoh, C., Mattrey, R. and Ridgway, S. 2004. Structural and functional imaging of bottlenose dolphin (*Tursiopstruncatus*) cranial anatomy. *Journal of Experimental Biology*. 207(21): 365–3665.

Islamy, R. A. and Hasan, V. 2020. Checklist of mangrove snails (Mollusca: Gastropoda) in South coast of Pamekasan, Madurai Island, East Java, Indonesia. *Biodiversitas* 21: 3127-3134.

International Whaling Commission. 1998 "*Special Permit whaling (also known as Scientific Whaling).*" https://iwc.int/permits.

International Whaling Commission, IWC, 2021 Report of the scientific committee. *Report of International Whale Commission* 48(53):118.

Iqbal, M., Setiawan, A. and Yustian, I. 2020. 'First inland record of bull shark Carcharhinus leucas (Carcharhiniformes: Carcharhinidae) in Indonesian Borneo,' *Ecologica Montenegrina*. doi: 10.37828/em.2019.24.9.

Jahved F. Maro, Agus Hartoko, Sutrisno Anggoro, Max R. Muskananfola, Erick Nugraha 2021a. Sea surface temperature and chlorophyll-a concentrations from MODIS satellite data and presence of cetaceans in Savu, Indonesia. AACL Bioflux, Volume 14, Issue 3. Pp: 1190-1200. http://www.bioflux.com.ro/aac

Jahved F. Maro, Agus Hartoko, Sutrisno Anggoro, Max R. Muskananfola, Lumban N. L. Toruan, Michael Dakahamapu, Erick Nugraha. 2021b. The appearance of the humpback whale (Megaptera novaeangliae) during the rainy season in the Alor

Waters, Indonesia. AACL Bioflux, Volume 14, Issue 3. Pp : 1596 -1608. http://www.bioflux.com.ro/aacl

James, P. S. B. R. 1974. 'An osteological study of the dugong Dugong dugon (Sirenia) from India', *Marine Biology*. doi: 10.1007/BF00389070.

Jefferson T. A., Leatherwood S., Webber M. A., 1993 FAO Species Identification Guide. *Mar Mamm World* 13(3):326–339.

Juraij. 2014. Seagrass Diversity for Dugong dugon Food Habbit. *Omni-Akuatika*.

Juraij, Bengen, D. G. and Kawaroe, M. 2014. 'Keanekaragaman jenis lamun sebagai sumber pakan dugong dugon pada desa busung bintan utara kepulauan riau,' *Omni-Akuatika*.

Karim, S., Kusuma, B. J. and Mahfud, T. 2019. 'Aplication the development of Balikpapan Bay Indonesia based on sustainable tourism,' *Geojournal of Tourism and Geosites*. doi: 10.30892/gtg.24103-340.

Kiszka, J. J. 2007. 'Atypical associations between dugongs (Dugong dugon) and dolphins in a tropical lagoon', *Journal of the Marine Biological Association of the United Kingdom*. doi: 10.1017/S0025315407055129.

Kügler, A., Lammers, M., Zang, E., Kaplan, M. and Mooney, T. 2020. Fluctuations in Hawaii's humpback whale Megaptera novaeangliae population inferred from male song chorusing off Maui. *Endangered Species Research* 43: 421-434.

Laran, S. and Gannier A. 2008. Spatial and temporal prediction of fin whale distribution in the northwestern Mediterranean Sea. *ICES Journal of Marine Science*. 65(7): 1260-1269.

Lee, P. B. and Nijman, V. 2015. 'Trade in dugong parts in Southern Bali,' in *Journal of the Marine Biological Association of the United Kingdom*. doi: 10.1017/S002531541500 1423.

Leeney, R. H., Berrow, S., McGrath, D., O'Brien, J., Cosgrove, R. and Godley, B. J. 2007. Effects of pingers on the behaviour of bottlenose dolphins. *Journal of the Marine Biological Association of the United Kingdom*. 87 (1) : 129–133.

Lipiński M., Vyver J. S. F., Shaw P., Sauer W. H. H., 2016 Life cycle of chokka-squid Loligo reynaudii in South African waters. *African J Mar Sci* 38(4):589–583.

Marsh, H., Harris, A. N. M. and Lawler, I. R. 1997. 'The sustainability of the indigenous dugong fishery in Torres Strait, Australia/Papua New Guinea,' *Conservation Biology*. doi: 10.1046/j.1523-1739.1997.95309.x.

McSweeney, D. J., Chu, K. C., Dolphin, W. F. and Guinee, L. N. 1989. North Pacific Humpback Whale Songs: a comparison of Southeast Alaskan feeding ground songs with Hawaiian wintering ground songs. *Marine Mammal Science*. 5(2): 139-184.

Mujiyanto M, Riswanto R, Dharmadi D, Ghiffary W. Composition and Distribution of Dolphin in Savu Sea National Marine Park, East Nusa Tenggara. *Indones Fish Res J*. 2018;23(2).

Muller PLS. 1775. Des Ritters Carl von Linné Königlich Schwedischen Leibarztes vollständigen Natursystems Supplements und Register-Band über alle sechs Theile oder Classen des Thierreichs. Mit einer ausführlichen Erklärung. Nebst drey Kupfertafeln [The Knight Carl von Linné Royal Swedish Personal Physician's Complete Natural System Supplements and Register Volume on All Six Parts or

Classes of the Animal Kingdom. With a detailed explanation. Along with three copper plates]. Nürnberg. Besch. Berlin. *Ges. Naturf.* 2: 584-592

Mohebbi-Kalkhoran, H., Zhu, C., Schinault, M. and Ratilal, P. 2019. Classifying Humpback Whale Calls to Song and Non-song Vocalizations using Bag of Words Descriptor on Acoustic Data. *18th IEEE International Conference on Machine Learning and Applications* (ICMLA): 865-870.

Nugraha, A. H, Ilham Antariksa Tasabaramo, Udhi Hernawan, Susi Rahmawati. 2019. 'Relationship of Distribution Seagrass Species with Dugong (*Dugong dugon*) Sighting at Liki Island-Papua,' *Omni-Akuatika*. doi: 10.20884/1.oa.2019.15.2.730.

Nybakken, J. W., 1992. *Biologi Laut. Suatu Pendekatan Ekologi* [*Marine Biology. An Ecological Approach*]. PT. Gramedia Pustaka Utama, Jakarta. hlm : 104-106.

Octavina, C, M Rizki Fazilah, Maria Ulfah, Syahrul Purnawan. 2020. Seagrass Biodiversity for Potential Feed of Dugong dugon at Lamteng Bay, ACEH. *Jurnal Ilmu dan Teknologi Kelautan Tropis*. doi: 10.29244/jitkt.v12i1.26747.

Palsbøll P. J., Allen J., Bérubé M., Clapham P. J., Feddersen T. P., Hammond P. S., Hudson R. R., Jørgensen H., Katona S., Larsen A. H., Larsen F., Lien J., Mattila D. K., Sigurjónsson J, Sears R., Smith T., Sponer R., Stevick P., Øien N., 1979 Genetic tagging of humpback whales. *Nature* 388:767-769.

Panigada, S., Zanardelli, M., MacKenzie, M., Donovan, C., Mélin, F. and Hammond, P. S. 2008. Modelling habitat preferences for fin whales and striped dolphins in the Pelagos Sanctuary (Western Mediterranean Sea) with physiographic and remote sensing variables. *Remote Sensing of Environment*. 112(8): 3400-3412.

Packard. T, Osma N, Fernández-Urruzola I, L. A. Codispoti L. A., Christensen J. P. & Gómez M. (2015). Peruvian upwelling plankton respiration: calculations of carbon flux, nutrient retention efficiency, and heterotrophic energy production. *Biogeosciences*. (12) 2641–2654.

Parsons, E. C. M., Wright, A. J. and Gore, M. A. 2008. The nature of humpback whale (Megaptera novaeangliae) song. *Journal of Marine Animals and Their Ecology*. 1(1): 21-30.

Praca, E., Gannier, A., Das, K. and Laran, S. 2009. Modelling the habitat suitability of cetaceans: Example of the sperm whale in the northwestern Mediterranean Sea. *Deep Sea Research Part I: Oceanographic Research Papers.* 56(4): 648-657.

Priosambodo, D., Nadiarti Nurdin, Khairul Amri, Yusran Nurdin Massa, Amrullah Saleh. 2007. Dugong Sighting at the Spermonde Islands South Sulawesi. *Jurnal Ilmu Kelautan Spermonde*. doi: 10.20956/jiks.v3i1.2118.

Putra, M. I. H., Lewis, S. A., Kurniasih, E. M., Prabuning, D. and Faiqoh, E. 2016. Plankton Biomass Models Based on GIS and Remote Sensing Technique for Predicting Marine Megafauna Hotspots in the Solor Waters. *IOP Conf. Series: Earth and Environmental Science* 47: 1-19.

Pringgenies D., Hartoko A., Maro J. F., Matsuishi T. 2021.The stomach content analysis of whales in Alor and Lamalera Islands, Savu The stomach content analysis of whales in Alor and Lamalera Islands, Savu Sea, Indonesia. *AACL Bioflux*, Volume 14, Issue 5. Pp : 3068-3075. http://www.bioflux.com.ro/aacl

Pringgenies D., Murdjani M., Sastrodihardjo S., Nganro N. R., 2000 Afirst attempt at culturing the squid Loligo duvaucelii orbigny, 1835 in Indonesia. *Phuket Mar Biol Cent Spec Publ* 21(1):187–188.

Rajamani, L. R. A. R. 2009. *The Conservation Biology of the dugong (Dugong dugon) and its seagrass habitat in Sabah, Malaysia: A basis for conservation planning*, Borneo Marine Research Institute.

Rako, N., Fortuna, C. M., Holcer, D. Š. Mackelworth, P., Nimak-Wood, M., Pleslic, G., Sebastianutto, L., Vilibic, I., Wiemann, A. and Picciulin, M. 2013. Lei- sure boating noise as a trigger for the displacement of the bottlenose dolphins of the Cres-LoŠinj archi- pelago (northern Adriatic Sea, Croatia). *Marine Pollution Bulletin*. 68 (1) : 77 – 84.

Records of yellow finned Barb *Mystacoleucus marginatus* (Valenciennes, 1842) in Brantas Basin, Indonesia. *Ecology, Environment & Conservation*. 26: 40-43.

Rodrigues, F. M., Silva, F. M. O., Trompieri-Silveira, A. C., Vergara-Parente, J. E., Miglino, M. A. and Guimarães, J. P. 2014. Morphology of the eyeball from the Humpback whale (Megaptera novaeangliae). *Microscopy Research and Technique*. 77(5): 348-55.

Rosas, C. L., Gil, M. N. and Uhart, M. M. 2012. Trace metal concentrations in Southern Right Whale (Eubalaena australis) at Península Valdés, Argentina. *Marine Pollution Bulletin*. 64(6): 1255-1260.

Savu, A., Mujiyanto, Riswanto, Nastiti, A. N. (2017) 'Efektifitas Sub Zona Perlindungan Setasea Di Kawasan Konservasi Perairan Tnp Laut Sawu, Nusa Tenggara Timur Effectiveness,' [Effectiveness of the Cetacean Protection Sub Zone in the Savu Sea Marine Protected Area, East Nusa Tenggara Effectiveness], *Coastal and Ocean Journal*.

Salim, D. 2011. Marine Mammals (Cetacean) at Savu Sea East Nusa Tenggara. *Jurnal Kelautan*. 4(1): 24-41.

Saptadjaja, AM., Hasan, V., Arief, M., Isroni, W. and Rozi. 2019. First record of threatened Asian Catfish, *Clarias batrachus* (Linnaeus, 1758) (Siluriformes, Clariidae) from Kangean Island, Indonesia. *Ecology, Environment and Conservation*. 26(3): 1055-1058. Savu, A., National, S.E.A., Park, M. and Tenggara, E. N. 2017. Effectivity of Protected Cetacean Sub-Zone at the Savu Sea Marine Protected Area, East Nusa Tenggara. *Coastal and Ocean Journal*. 1: 1-12.

Santos MB., Pierce GJ., Herman J., López A., Guerra A., Mente E., Clarke MR., 2001 Feeding ecology of Cuvier's beaked whale (Ziphius cavirostris): A review with new information on the diet of this species. *J Mar Biol Assoc United Kingdom* 81(4):687–694.

Secchi, E. R., Rosa, L. D., Kinas, P. G., Nicolette, R. F., Rufino, A. M. N. and Azevedo, A. F. 2011. Encounter rates and abundance of humpback whales (Megaptera novaeangliae) in Gerlache and Bransfield Straits, Antarctic Peninsula. *J. Cetacean Res. Manage*. (Special Issue) 3: 107-111.

Sediadi A. Effek Upwelling Terhadap Kelimpahan Dan Distribusi Fitoplankton Di Perairan Laut Banda Dan Sekitarnya. *MAKARA Sci Ser*. 2010;8(2).

Sheppard, J. K, Anthony R Preen, Helene Marsh, Ivan R Lawler, Scott D Whiting, Rhondda E Jones. 2006. 'Movement heterogeneity of dugongs, Dugong dugon (Müller), over large spatial scales,' *Journal of Experimental Marine Biology and Ecology*. doi: 10.1016/j.jembe.2006.01.011.

Sheppard, J. K, Anthony. 2010. 'Dugong habitat use in relation to seagrass nutrients, tides, and diel cycles,' *Marine Mammal Science*. doi: 10.1111/j.1748-7692.2010.00374.x.

Sydeman WJ., García-Reyes M., Schoeman DS., Rykaczewski RR., Thompson SA., Black BA. and Bograd SJ. 2014. Climate change and wind intensi- fication in coastal upwelling ecosystems. *Science*. 345 (6192) : 77–80.

Tsutsumi, C, Kotaro Ichikawa, Nobuaki Arai, Tomonari Akamatsu, Tomio Shinke, Takeshi Hara, Kanjana Adulyanukosol. 2006. 'Feeding behavior of wild dugongs monitored by a passive acoustical method,' *The Journal of the Acoustical Society of America*. doi: 10.1121/1.2221529.

Tynan, C. T., Ainley, D. G., Barth, J. A., Cowles, T. J., Pierce, S. D. and Spear, L. B. 2005. Cetacean distributions relative to ocean processes in the northern California Current System. *Deep Sea Research Part II: Topical Studies in Oceanography*. 52(1): 145-167.

Werth, A. J., Kosma, M. M., Chenoweth, E. M. and Straley, J. M. 2019. New views of humpback whale flow dynamics and oral morphology during prey engulfment. *Marine Mammal Science*. 35(4).

Widodo, M. S., Hasan, V., Mukti, A. T. and Kusuma, B. 2020. Distribution of Dwarf Snakehead *Channa gachua* Hamilton, 1822 (Teleostei, Channidae) on Brantas River Basin, Indonesia. *Ecology, Environment and Conservation*. 26 (2) : 618-621.

Wirasatriya, A., Setiawan, J. D., Sugianto, D. N., Rosyadi, I. A., Haryadi, H., Winarso, G., Setiawan, R. Y. and Susanto, R. D. 2020. Ekman dynamics variability along the southern coast of Java revealed by satellite data. *International Journal of Remote Sensing*. 41(21): 8475–8496.

Woodward, B. L., Winn, J. P. and Fish, F. E. 2006. Morphological specializations of baleen whales associated with hydrodynamic performance and ecological niche. *Journal of Morphology*. 267(11): 1284-94.

Valen, F. S., Soemarno Widodo, M. S., Wiadnya, D. G. R. and Hasan, V. 2019. *Contemporary distribution*.

World Wildlife Fund, (2020). Mamalia Laut. World Wildlife Fund. *Laporan Survei Mamalia Laut* [*Marine Mammal Survey Report*].

Yusmalinda NLA, Anggoro AW, Suhendro DM, Ratha IMJ, Suprapti D, Kreb D, et al. Identifikasi Jenis Pada Kejadian Cetacea Terdampar Di Indonesia Dengan Teknik Molekuler [Identification of Species in Cetacean Stranding in Indonesia Using Molecular Techniques]. *J Ilmu dan Teknol Kelaut Trop*. 2018;9 (2).

Zerbini, A., Andriolo, A., da Mata, J., Simões, P., Siciliano, S., Pizzorno, J., Waite, J., DeMaster, D., and VanBlaricom, G. 2004. Winter distribution and abundance of humpback whales (Megaptera novaeangliae) off Northeastern Brazil. *Journal of Cetacean Research and Management*. 6(1): 101-107.novaeangliae) off northeastern Brazil. *Journal of Cetacean Research and Management*.

Biographical Sketch

Prof. Dr. Ir. Agus Hartoko, MSc

Affiliation: Dept of Aquatic Resource Management. Faculty of Fisheries and Marine Science. Diponegoro University. Semarang Indonesia

Education: PhD

Business Address: Dept. of Aquatic Resource Management. Faculty of Fisheries and Marine Science. Diponegoro University. Jl. Prof Jacub Rais. Semarang Indonesia

Research and Professional Experience:

1982-1983	Team Studi Dampak Sosial-ekonomi Radius 30 dan 60 Km Rencana PLTN Ujung Watu Jepara. UNDIP – BATAN- CESEN Italy
1982	Team Study on the Animal Associated With Branching Corals at Jepara Waters. UNDIP - New Castle Univ
1982-1986	Team Development of Marine Science Study Programms at Six Universities in Indonesia. Ministry of Education & ADB
2004	Team Leader Supply of Coastal Resource Thematic Datasets for West Sumatra, MCRMP- Ministry for Marine and Fisheries & ADB Team Leader Detail Mapping for Oil Pipe of Blora – Tuban. Exxon Mobil
2021	Team Leader "Potential Placement of Fish Aggregating Device/ FADs (Rumpon) in Indonesia Fisheries Management Areas (WPP NRI)." Ministry of Marine and Fisheries- BAPPENAS - UNDP

Professional Appointments:

Guest Lecturer – Faculty of Fisheries Sciences. Hokkaido University. Japan, 2019.

Honors:

UNEPNairobi Kenya - 2010, IOC-UNESCO 2013, United Nations Development Programms-2021.

Publications from the Last 3 Years:

1. Delianis Pringgenies, Agus Hartoko, Jahved Ferianto Maro, Takashi Matsuishi, 2021. The stomach content analysis of whales in Alor and Lamalera Islands, Savu Sea, Indonesia. *AACL Bioflux,* 2021, Volume 14, Issue 5: 3068-3075. http://www.bioflux.com.ro/aacl.
2. Jahved F. Maro, Agus Hartoko, Sutrisno Anggoro, Max R. Muskananfola and Andri P. Timung.2021. Dolphin Appearance in the Alor - Pantar Strait with a Variable Oceanographic Approach. *Eco. Env. & Cons.* 27 (November Suppl. Issue): 2021; pp. (S127-S131).Copyright@ EM International. ISSN 0971–765X. PP: 127-131.
3. Jahved F. Maro, Agus Hartoko, Sutrisno Anggoro, Max R. Muskananfola, Erick Nugraha. 2021. Sea surface temperature and chlorophyll-a concentrations from MODIS satellite data and presence of cetaceans in Savu, Indonesia. *AACL Bioflux*, 2021, Volume 14, Issue 3. http://www.bioflux.com.ro/aacl. PP: 1190-1200.
4. Jahved F. Maro, Agus Hartoko, Sutrisno Anggoro, Max R. Muskananfola, Lumban N. L. Toruan, Micael Dakahamapu, Erick Nugraha. 2021. The appearance of the humpback whale (*Megaptera novaeangliae*) during the rainy season in the Alor Waters, Indonesia. *AACL Bioflux*, 2021, Volume 14, Issue 3. http://www.bioflux.com.ro/aac. PP: 1596-1608.
5. Hartoko A, Jahved Ferianto Maro, Delianis Pringgenies (in press 2022). Cold Seawater, Mass Fish Death Phnomena and Dolphin Appearance at Pantar Strait Savu Sea East Nusa Tenggara Indonesia. Submitted to *International Journal of Environmental Science and Development*.
6. Helmi. M, Agus Hartoko, Hifzhan Husna, Mochamad Indra Bayu Ardiansyah. 2022. Integration of Sentinel-1A SAR, Hydrodynamic and Trajectory Model of Oil Spill at Balongan Coastal Water Java Sea Indonesia. *Geographia Technica*. Vol .17 (1) : 58-76. DOI : 10.21163/GT_2022.171.05. ISBN (Print): 1842-5135; ISBN (Online): 2065-4421. Https://technicalgeography.org.
7. Hartoko.A, Delianis Pringgenies, Amelia. 2020. Morphology and Molecular Biology of Benthic Java Sea Shark Ray *Rhina ancylostoma* Bloch and Scheider 1801 (Elasmobranchia: Rhinidae). *Annual Research and Review in Biology*. 35(4) : 19-31. DOI:10.9734/ARRB/2020/v35i4 30208.

8. Jahved Ferianto Maro, Agus Hartoko, Sutrisno Anggoro, Max Rudolf Muskananfola· *New Discovery of the Friendly Behavior of Dugong (Dugong-dugon) (Muller, 1776) (Sirenia: Dugongidae) towards Humans in the Waters of Kabola, Alor, Indonesia.*

Chapter 4

An Assessment of Bycatch and Discard in Benthic Trawling in Northern Tunisia (GFCM- GSA 12), Western Mediterranean Sea

Sami Mili[1,2]
Rym Ennouri[1,3]
Manel Fatnassi[1,*]
Abdelhamid Helali[1]
Faten Alayete[1]
and Hajer Zarouk[1]

[1]University of Carthage, Higher Institute of Fisheries and Aquaculture of Bizerte, Errimel, Bizerte, Tunisia
[2]Laboratory of Fisheries Science, National Institute of Marine Sciences and Technologies, Salammbô, Tunis, Tunisia
[3]Laboratory of Ecology, Biology and Physiology of Aquatic Organisms, Faculty of Sciences of Tunis, University of Tunis El Manar, Tunis, Tunisia

Abstract

The release of marine organisms at sea following each fishing operation is a common global practice, and the Mediterranean region is no exception. There are several reasons for discards: some species are not marketable, others do not find a buyer on the market, or do not reach the legal size for marketing. To gain insight into the impact of benthic trawling both on the marine ecosystems and on the exploitable fishery

---

* Corresponding Author's Email: manel.ft@gmail.com.

In: Advances in Marine Biology. Volume 6
Editors: Adam Kovács and Patrik Nagy
ISBN: 979-8-88697-996-1

resources of the northern region of Tunisia (GSA12), an observer went on fishing units to collect the data needed for discard analysis.

A total of four fishing surveys were conducted during the year 2022 onboard benthic trawlers. The quantity of unwanted marine organisms tossed back into the sea during these trips was estimated at 301.45 kg, which corresponds to an overall hourly discharge rate of 4.63 kg/h. The highest species discard was recorded at depths greater than 200m. During the winter and spring periods (January-April), the total amount of bycatch reached 21% of the production. Furthermore, the results showed that the levels of discards varied according to the species group: fish, crustaceans, and cephalopods. The diversity of marine organisms thrown back is also influenced by the sea depth and condition, as well as the fishing area. The ecological indices analysis indicated that the status in some areas of GSA12 can be described as ecologically poor with a tendency towards equilibrium.

Studies related to the assessment of exploited stocks must consider bycatch. The reason for this is that the quantities discarded are a significant component that must be estimated and included in the fisheries management models.

Keywords: northern Tunisia, benthic species, GFCM-GSA12, bycatch, biodiversity

Introduction

Tunisia has more than 1,300 kilometers of coastline. The Tunisian marine space is divided into three distinct zones: north, east, and south. The northern region (GSA 12) starts at the Algerian-Tunisian border and extends to Ras Mustapha (Kelibia). The Eastern region (Chebba) is made up of Ras Mustapha to Ras Kapudia. The southern region (Gulf of Gabes), which is formed by a large continental shelf, stretches from Ras Kapudia to the Libyan border (Jort, 1995; Jarboui et al., 2005; Rjeibi, 2012).

Fishing is practiced all along the Tunisian coast. It is dependent on a network of fishing ports located along the coast. There are currently 41 fishing ports, which can be divided according to their importance into deep-sea ports (10) that accommodate large units (trawlers, seiners), and coastal ports (31) that are home to small and medium-sized vessels. This fleet is at the heart of the growth of the local fishing economy.

The fishing industry is vital to the Tunisian economy. Recent statistics indicate that this activity generated a total production of 137.470 tons and

144.8 million TND in 2020, alone. As a result, it ranks second in terms of agricultural exports. This production has been ensured by a diverse fishing fleet consisting primarily of small-scale fishing boats, seiners, trawlers, and various other fishing methods such as fishing on foot (Jarboui et al., 2005; Ennouri et al., 2015). In Tunisia, a broad range of fishing methods and gear are used. Those deployed in small-scale (coastal) fisheries vary by region and season, targeting numerous species. We primarily come across nets (gillnets, trammel nets), longlines, traps, etc. (Ennouri et al., 2015). Purse seines, either rotating or sliding, are used to catch pelagic species. Trawling is also carried out by vessels based in a variety of ports, using shrimp trawls, Mediterranean trawls, Italian trawls, and large vertical opening trawls (GOV) (M'rabet et al., 2003; Bdioui & Mrabet, 2009). In general, tides in the North range from three to five days, with a haul duration of two to three hours. In the east, they last four to five days with an average haul duration of two hours, and in the south, six to eight days with a haul duration, generally, of one to two hours (Jarboui et al., 2005).

The benthic trawl fishery produces discards that include marketable species and diverse benthic fauna. These organisms are discarded for regulatory or commercial reasons and are mostly immature specimens. Bycatch and discard are made up of various ecological groups.

Discarding represents an economic and ecological waste of exploitable fishery resources. This issue, however, continues to be a major concern in Mediterranean fisheries. However, no previous investigation into the discard and by-catch of Tunisia's northern region (GSA 12) in the western Mediterranean Sea has been realized till date. In this work and in an effort to attenuate the damages caused by discard in this fishing zone, we attempt to determine the discard's types and characteristics as caught by benthic trawlers. We used a direct sampling method during fishing operations by going on board a professional fishing trawler operating in this area. The sampling was followed by work in the laboratory to treat the bycatch samples, including species' identification, estimation of quantities, measurements, etc.

Overview

Presentation of Northern Tunisia (GFCM- GSA 12)

Geographical Position

Tunisia is a northern African country with a strategic location in the Mediterranean, attracting migratory species and facilitating maritime navigation. When the lengths of its various islands' coastlines are added up, the total length reaches 2.290km (Gipp, 2019).

The northern zone has a 400-kilometer coastline. The coasts occupy a continental terrain dominated by small mountains and, in some places, open hills (Rjeibi, 2012). It is divided into two sections. The first area has a dense hydrographic network and is located between the Tunisian-Algerian border and Cap Blanc. The turbulence of the sea, as well as the frequent violent winds, define the shoreline. The second sector is characterized by extremely powerful waves that weaken and erode the cliffs, resulting in extremely uneven relief. This is because the area is exposed to a deep sea and the continental shelf is narrow, rocky, and steep. These generally rocky coasts are dotted by small mobile sandy beaches and dunes (Ben Ismail et al., 2010; Amrouni et al., 2014; Ben Ismail, 2014; Khoufi, 2015).

Topographical Conditions

The area between Cap Blanc and Kelibia is distinguished by a unique fauna. The pressures of urbanization and the development of tourist activities are particularly intense in this region. The water is shallow, and the hydrographic network is extremely vital (Amrouni et al., 2014). In this shallow zone, marine invasions formed large lakes and gulfs which shelter numerous beaches. In this area, the terrain is more aerated, and the coastline is soft, with sandy beaches (Rjeibi, 2012; Khoufi, 2015).

This sector (GSA 12) is distinguished by the presence of several Capes: *Cap Negro, Cap Zbib, Cap Blanc, Cap Chauve, Cap Farina, Cap Serrat,* and *Cap Engela*. The presence of two archipelagos, golf, and two lagoons creates distinct ecosystems (Amrouni et al., 2014).

The northern Tunisian coast (GSA12) has a rocky shoreline that follows the country's exceptionally undulating mountainous topography. It has rough cliffs made of limestone and sandstone that have been eroded by the waves. There are few narrow beaches along the rivers and estuaries of alluvial plains, with a small area of shallow water between the beach and the open sea (Rjeibi, 2012; Chemmam-Abdelkader et al., 2003).

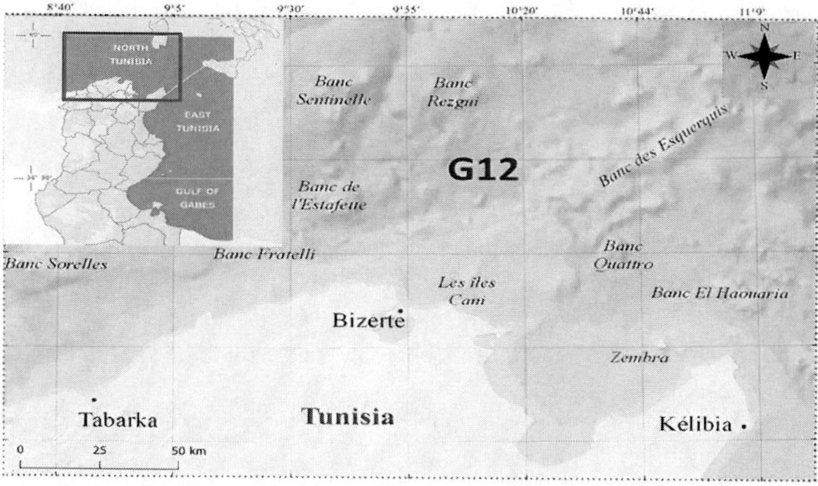

Figure 1. Geographical location of Tunisia's northern region (GFCM-GSA12).

Climatic Conditions

The northern Tunisian coast, like the rest of North Africa and Western Mediterranean countries, is endowed with a Mediterranean climate with long periods of good weather; it has an arid climate with little rainfall and is influenced by disturbances from the Balearic Islands and the Gulf of Genoa (Rjeibi, 2012). The hot season that follows the summer is dry, while the winter is marked by irregular rain falls in the form of violent showers of short duration (Ben Ismail, 2014; Khoufi, 2015).

Bathymetry

In the north of Tunisia, the continental shelf presents a variable extent and depth. In the western and central parts of this area (Tabarka-Bizerte), the shelf reaches a depth of 200m. The slope is located very close to the coast. While in the eastern region, from Bizerte to Cap Bon, the continental shelf is wider, with a lower limit not exceeding 100 m. It reaches a width of 20 nautical miles, due to the slope (Rjeibi, 2012).

Hydrology

Temperature

Temperature is a critical regulator for marine life. It influences the migratory behavior and reproduction of marine organisms on a spatial and temporal scale (Rjeibi, 2012; Ben Ismail, 2014).

Temperature, in the western sector in the north of Tunisia, varies considerably depending on the season and depth. Winter and spring temperatures range between 14.3°C at the surface and 50m depth, 14.4°C at 100m, and 13.5°C at 150m. A thermocline appears between 20 and 50m depth in the summer. Temperatures range from 22°C to 23°C depending on the station (Rjeibi, 2012; Ben Ismail, 2014). In the Galite region, the temperature decreases sharply with depth to reach 13°C, but it remains constant beyond 300m of depth (Rjeibi, 2012).

Salinity

Salinity is a chemical parameter that has a significant impact on oceanic processes. It regulates the density of seawater along with temperature and pressure. These density changes induce thermo-haline circulation, which influences climate regulation. The Mediterranean is known for high evaporation (Lynne et al., 2011).

During the winter, the Atlantic surface waters in the Galite area have a salinity of 36.73‰. The latter reaches 37.5‰ at a depth of 50 to 150 meters. The greater salinity is induced in deep waters, and more water tends to flow west (between 38‰ and 38.5‰) (Ben Ismail, 2014; Würtz, 2010).

Currentology

Herman (2007) denotes two masses of water flowing in opposite directions in the Mediterranean: an intermediate-mass (Levantine Water) moving east to west and another surface mass (Atlantic Water) flowing west to east. The Atlantic Current (MAW) deviates from the coast in the Tunisian sector, particularly in front of the Galite and Esquerquis areas, which are the sites of eddy movements. The MAW splits into two branches in the Sicilian Channel. One travels east along northern Sicily and then north along the west coast of Italy. The second enters the Sicilian Channel and takes up the entire top layer (Sammari & Brahim, 1996). During winter convection in the Eastern Mediterranean, the Levantine Current (LW) forms in the Levantine Sea. It enters the Western Mediterranean via the Siculo-Tunisian Strait, then makes a cyclonic journey along the coast in depths ranging from -300m to -800m. Astraldi et al. (2002a) demonstrated that, in addition to the (LW) and (MAW), other deep-water masses participate in the exchanges in the Central Mediterranean, particularly between the Siculo-Tunisian Channel and the Tuniso-Sardinian Channel. These water masses' movements vary seasonally due to meteorological conditions and the nature of local reliefs (Euromodel Group, 1995; Roussenov et al., 1995; Alpert et al., 1996). Indeed, a homogenization of the water mass in the study area (GSA12) can be observed, owing to the strong surface or bottom water currents (Sammari & Gana, 1995).

Trawlable Bottoms

The northern area of Tunisia (GSA12) is divided into sectors based on underwater morphology. Trawling is possible in the Canal, the Galite, and Cap Serrat from -130 to -450m in the western part. With the exception of the Estafette area, fishing is feasible at the bathyal level from a depth of -300m to -650m in the central part. Apart from the Esquerquis area, all circalittoral and bathyal bottoms (-50 to -450m) are exploitable in the eastern part, particularly in the Gulf of Tunis. Fishing can be done in depths below -150 meters around Cani Island (Northeast of Cap Zebib). The Cap Bon area is less suitable for fishing because the underwater relief is rugged, the continental shelf is limited, and deep channels cut the steep slope (Chemmam-Abdelkader et al., 2003; Sammari & Gana, 1995; Cherif, 2014).

Trophic Richness

According to Ben Maiz (1995), Boudouresque (1997), and Bradai et al. (2004) in Tunisia, 414 taxa with 333 fish species have been identified. The majority of these species, like the Mediterranean ichthyofauna, are native to the Atlanto-Mediterranean region (Quignard, 1978). The northern coast of Tunisia (GSA12) is home to vast meadows of Posidonia, which serve as spawning grounds for a variety of fish, cephalopods, and benthic invertebrates such as sponges, bryozoans, and annelids. The abundance of algae in the circalittoral bottoms attests to the environment's nutritive richness (Ben Mustapha et al., 1999).

Fisheries Exploitation in the Northern Region of Tunisia (GSA12)

The rationalization of resource exploitation, particularly the fishing techniques and gear used, is the fundamental basis of sustainable fishery management

(Jabeur, 1999; Mili et al., 2021). In Tunisia, the fishing industry plays an important role in the national economy, providing food, employment, recreation, trade, and social well-being for current and future generations alike. This sector contributes to the economic dynamization, marketing, transformation, and conservation of seafood products. Eventually, this activity ranks second in terms of exports after olive oil. Furthermore, fishing provides both direct and indirect employment (4% of the Tunisian labor force in 2020 (DGPA, 2020)). On the other hand, the consumption of fish per capita is currently 20.5kg, with an unbalanced regional distribution. The consumption in the interior regions does not exceed 1.5kg per capita per year (FAO, 2023).

According to the Tunisian General Directorate of Fisheries and Aquaculture Statistics (DGPA, 2020), there has been a small variation in production over the last 15 years ($p > 0.05$). In 2020, the landings totaled 1374.3 tons, worth 1.448.000 Tunisian dinars.

Tunisia's fishing sector development is based on the rational management of fishery resources and the preservation of a balance between fishing efforts and exploitable resources in various regions. This is possible through conducting fishing surveys and monitoring fleet evolution (Mili et al., 2021). On this basis, fishing efforts will be focused primarily on underexploited areas, such as Tunisia's north (GSA12) and offshore areas. This act must be linked to an increase in the exploitation rate of the species through the use of a specialized fleet equipped with modern methods and gear that allow for fish handling and processing on board (Anonyme, 2002).

In the following sections, we describe the characteristics of the operating fisheries in northern Tunisia (GSA12), including the fleet, the maritime population, and the fishing techniques used. It is important to note that the majority of the data are based on the statistics provided by the Tunisian General Directorate of Fisheries and Aquaculture (DGPA, 2020).

Fishery Production in the GSA12

The delegations that constitute GSA12 are composed of several important fishing towns in terms of port infrastructure (Rjeibi, 2012; Romdhane, 1993; Abdesselem, 1998). Tabarka, Bizerte, La Goulette, and Kelibia have deep-sea ports, and Ghar El Meleh, La Galite, and Sidi Daoud have coastal fishing ports. In addition to this infrastructure, there are several landing points scattered along GSA12, particularly in Bizerte and Tunis delegations (Abdesselem, 1998).

Total production (all species) in Tunisia's northern region (GSA12) is estimated to be 25252.27 tons between 2006 and 2020. This production is derived from three major fishing activities (small-scale fishing, trawling, and seining), which account for 22% of the total catch in Tunisia (DGPA, 2020). During the period 2006-2020, we observed a slight decrease in landings in Tunisia's northern region ($p<0.05$), from 30.000 tons in 2006 to 23.000 tons in 2020.

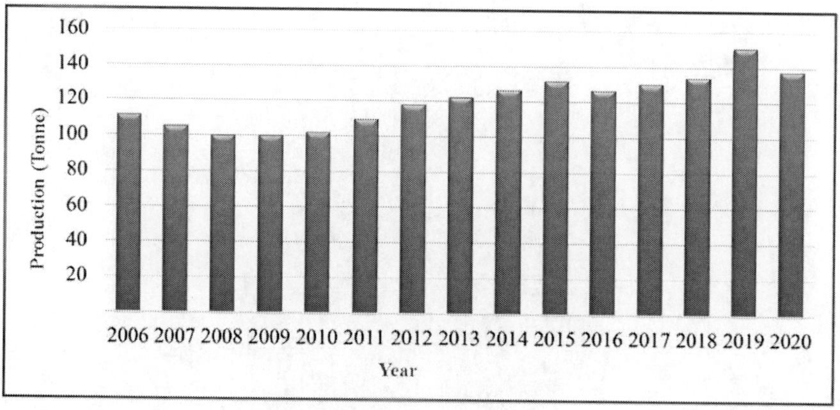

Figure 2. Evolution of landings in Tunisia (DGPA, 2006-2020).

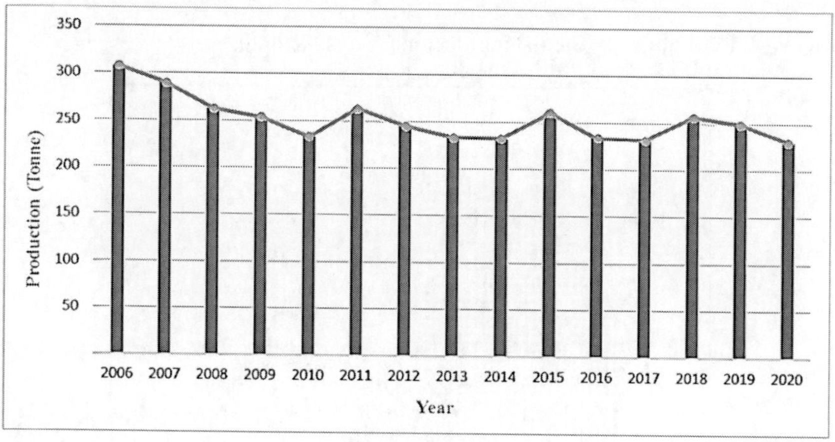

Figure 3. Evolution of fisheries' production in the northern region of Tunisia (GSA12) (2006-2020).

Fishing Effort

The fishing fleet in GSA12 consists of a diverse range of vessels, from simple traditional boats to trawlers and seiners. This fleet has grown tremendously and has become more and more powerful with 3.488 active units in 2020. It is primarily composed of small-scale fishing vessels (88%). The majority of these boats (52%) are non-motorized. Trawlers and seiners account for 5% and 3% of the total number, respectively. Boats working in reservoirs, lagoons, and aquaculture activities make up the remaining 4%.

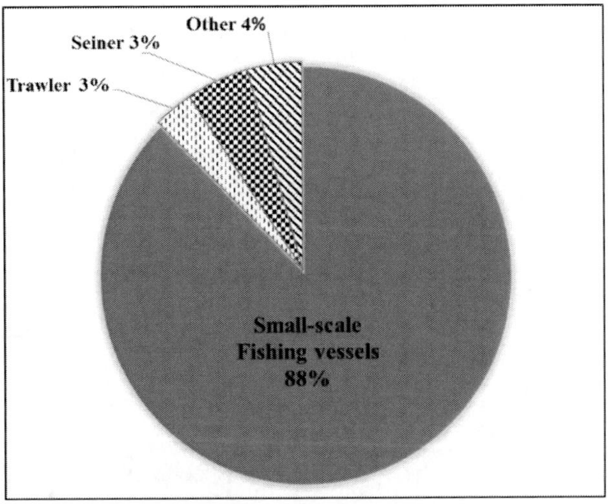

Figure 4. Distribution of the fishing fleet in GSA12 in 2020.

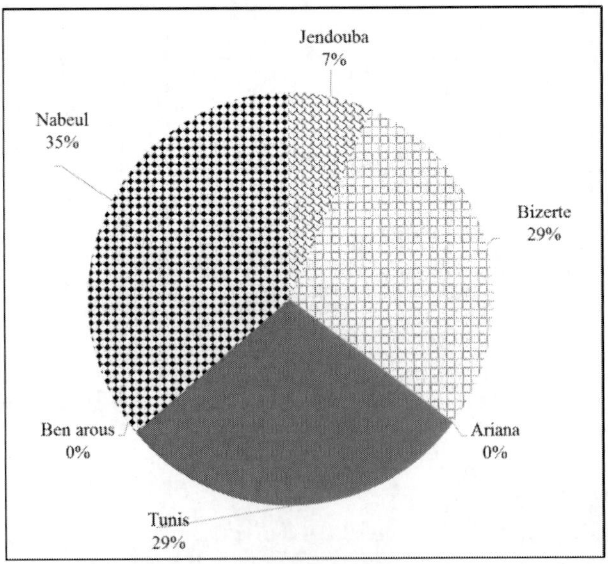

Figure 5. Distribution of the fleet in GSA12 by delegation in 2020.

The governorate of Bizerte governs most of the regional fleet (41%). Ben Arous Delegation comes in second with 18%. The rest of the regions share the following percentages: Ariana (8%), Tunis (11%), Nabeul (14%), and Beja/Jendouba (8%).

Workforce

In 2020, fishing was carried out by 15.179 fishermen in GSA12 (DGPA, 2020). However, the distribution of this region's active population is unbalanced in terms of delegations and fishing activity. Indeed, the delegations of Bizerte and Ben Arous account for 57% of the workforce in the GSA12 Region. The other 43% of the maritime population spread within the delegations of Nabeul, Tunis, Ariana, and Jendouba.

Table 1. Active maritime population in northern Tunisia (GSA12) in 2020

Activity	Jendouba	Bizerte	Ariana	Tunis	Ben Arous	Nabeul
Small-scale fishing vessels	544	4052	1080	1058	2248	1738
Trawler	90	288	-	338	300	300
Seiner	36	582	-	54	1020	879
Other	200	127	-	87	158	0
Total	870	5049	1080	1537	3726	2917

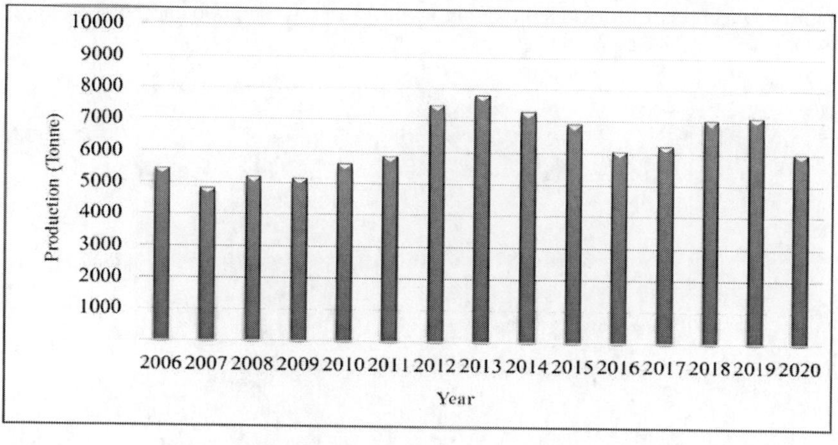

Figure 6. Trawl production in the GSA12 region between 2006 and 2020.

Benthic Trawling

Production

The monitoring of landings from bottom trawl catches in the GSA12 Region reveals big fluctuations in production. During the period 2006-2020, it ranged from 5.000 to 8.000 tons (Figure 6). Following an increase between 2006 and 2011, trawl production continued to rise, attaining 7.800 tons between 2012 and 2014. However, it dropped in 2016 to 6.000 tons, and then climbed again to 7.000 tons in 2019.

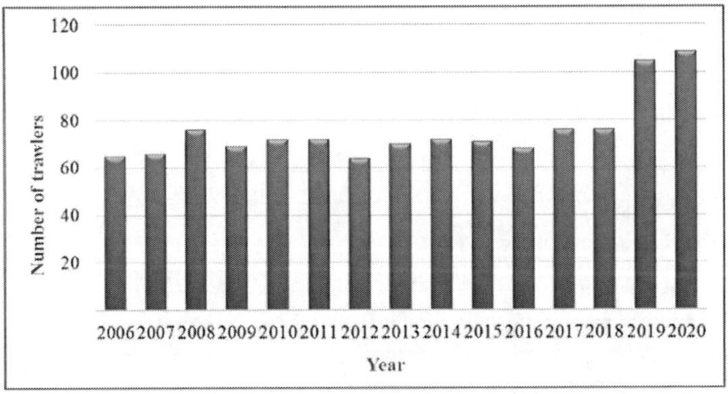

Figure 7. Number of trawlers operating in GSA12 between 2006 and 2020.

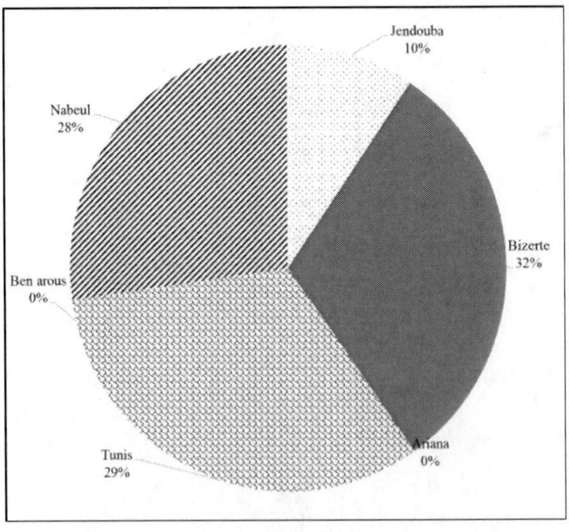

Figure 8. Distribution of trawlers by delegation in the GSA12 Region (2006-2020).

This fluctuation can be attributed to the recent increase in the number of trawlers. Furthermore, this fleet has undergone significant technological evolution, allowing for better navigation in adverse weather conditions and enhanced fish detection performance.

Fishing Fleet

An assessment of the GSA12 fishing fleet demonstrated a significant rise in the number of vessels, from 65 in 2006 to 109 in 2020. This variation is mainly attributable to the increase in freezer trawler units targeting deep-sea crustaceans.

Additionally, we observed that the governorate of Bizerte has the highest number of vessels by delegation (32%). Tunis and Nabeul delegations occupy the second rank with 58% of the total number. The regions of Beja/Jendouba and Ben Arous/Ariana share the remaining 10%, respectively.

Workforce

In 2006, in Tunisia's northern region (GSA12), there were 750 sailors engaged in benthic trawling with a projected increase to 1300 by 2020. An important increase was recorded in 2019, due to a rise in the number of trawlers (44). The regulations, imposed during the COVID-19 pandemic, did not allow for more than five fishermen per trawler, which explains the drop in the workforce that took place in 2020.

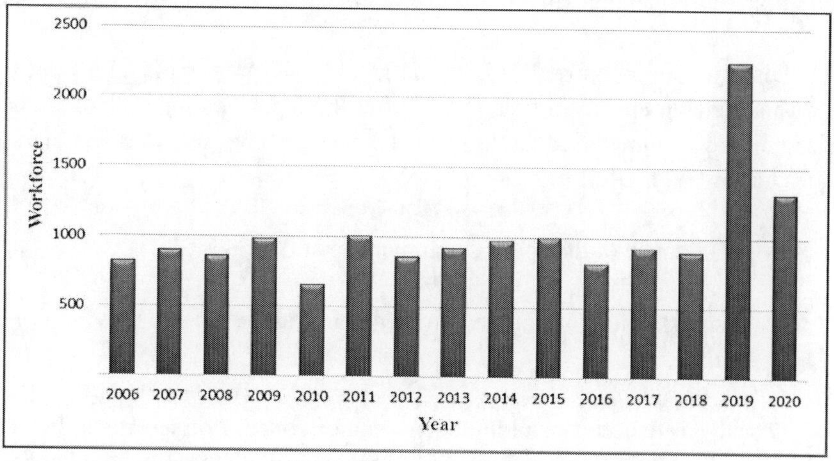

Figure 9. Evolution of the active trawling workforce operating in the GSA12 region (2006-2020).

The distribution of active trawling workforce shows that Tunis, Bizerte, and Nabeul delegations are at the top. These fishing districts employ 93% of Tunisia's active trawling workforce.

Bycatch and Discard from Fishing

Fishing bycatch and discard refer to any unwanted marine organisms captured, but not landed (Morizur et al., 1997). Every fishing operation includes the discarding of marine organisms, mostly non-commercial species. These are thrown back into the sea either because they are too small (juveniles), inedible (benthos), or damaged. Discards represent mortality caused by fishing gear and must be considered in stock assessments to avoid bias in predictions. Discards are important in multi-species assessments (Morizur et al., 1997).

One of the goals of research into the efficiency of fishing techniques is to reduce discards. Discarded small-sized species, as well as benthic organisms that play an important role in the marine ecosystem, are part of the stock.

Origin of Fishing Discards and Bycatch

Discard rates vary depending on the fishing method used in shallow and deep water. The characteristics of the fishing gear (mesh size, selectivity), the maneuvering (trawling), the fishing area (sandy or seagrass bottom), and the duration of immersion all contribute to this phenomenon (Morizur et al., 1997).

Discards are composed of both marketable and non-marketable species. Non-marketable species are catches below the legal landing size or market size, injured specimens, or catches above the quota. The quantities and discard rates depend on the following:

- abundance of the species and recruitment;
- gear selectivity;
- fishing techniques and commercial size; and
- quota.

Discard in fisheries has long been a major concern of scientists. Indeed, the resource exploitation rate cannot be estimated solely by analyzing landings but rather by evaluating all types of catches. Information on the small caught specimens can also provide valuable indicators of the strength or weaknesses of recruitment (Guérineau et al., 2009).

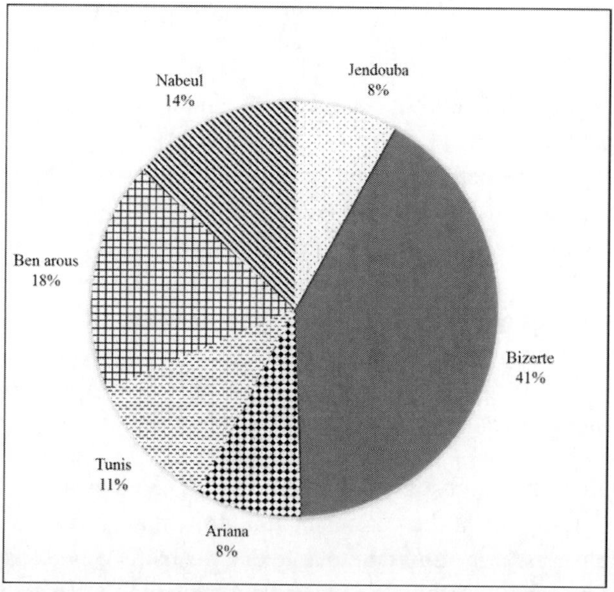

Figure 10. Distribution of the active marine trawler workforce by delegation in GSA12 (2006-2020).

Material and Methods

To study discards, sampling can be done in different ways, including on-board observation and sampling by fishermen. In this study, we opted for the most effective method: on-board observation. This technique allows for direct sampling of landings and discards during fishing operations. However, due to the reluctance of some fishing skippers, it was not easy to have an observer on board professional fishing units. Furthermore, performing this task can be challenging due to the difficult working conditions on professional trawlers.

To assess discards and bycatch in the GSA12 Area, we examined the specific compositions of the discarded small fish and benthos both qualitatively and quantitatively on a monthly basis between January and April 2022. Sampling was made on board professional trawlers with comparable technical characteristics: engine power from 780 to 900 hp and a total length ranging from 28 to 32m.

We deployed the four-sided trawl, which is a popular fishing gear among Tunisian offshore trawlers. The tides lasted 3 to 5 days, with a haul duration ranging from 3 to 5 hours. The sampling unit used is one haul and the following elements were recorded for each trawling operation:

- geographical coordinates of the beginning and end of each fishing operation;
- depth at the beginning and end of each fishing operation;
- start and end-time of each haul;
- trawling speed and direction of each haul;
- specific composition of the catch and benthos; and
- quantities of commercial species and discards.

Presentation of the Fishing Area

The sampling operations were carried out in Tunisia's northern region (GSA12). The depths of the various benthic trawling operations ranged from 150 to 400 meters. It is important to note that the hauls were carried out at the decision of the boat's captain, and we had no say in the choice of their location. This provided us with authentic information relevant to the reality of the professional activity. The fishing operations were carried out between January and April 2022, as shown on the map below.

Material (Fishing Gear)

Trawler units in the GSA12 Region generally deploy the benthic trawl for fishing. In the northern region of Tunisia, two types of gear are commonly used: shrimp trawls and Mediterranean trawls (Italian). This trawl's mesh size is relatively small, ranging from 45mm (wings) to 24mm (cod-end). The trawler is classified as harmful and non-selective fishing gear in the absence of adequate cutting and rigging (M'rabet et al., 2003).

Methodology

Fieldwork

We conducted a survey on the board of professional trawlers to study discards quantitatively and qualitatively. All trawl haul data, including date, time, haul number, geographical position, and depth, were collected and recorded. Following the sorting of commercial species in each fishing operation, a representative sample of the discarded species was taken randomly. At the end of each survey, samples were taken to the laboratory for analysis. The 24 trawling operations were carried out during the study period (January to April 2022) and at different bathymetries (150-400m).

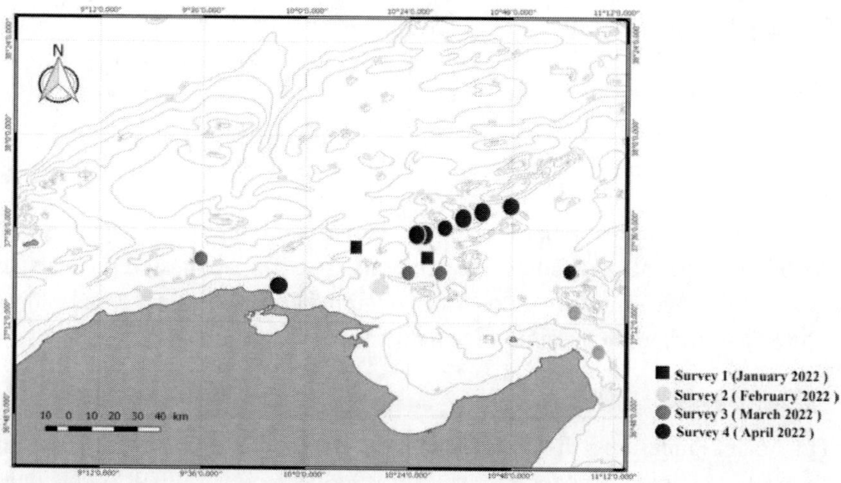

Figure 11. Location of trawl hauls in GSA12 (January - April 2022).

Laboratory Work

In the laboratory, we analyzed and processed the discarded samples taken on board the trawlers. For each sample, we identified the species caught and the benthos and sorted them following standard keys, such as Ridel (1959) and Fisher et al. (1987). For each collected specimen, we performed the following operations: measurement of the total length (TL) to the nearest mm and weighing with a precision of 0.01g. Regarding cephalopods, we measured the dorsal length of the mantle.

Additionally, we calculated the indices related to biodiversity.

Index of Abundance

The following formula was used to calculate the abundance indices (hourly yield) for commercial species and discards of small fish and other benthic species (Grall & Coïc, 2006):

$$Rhi(Kg/h) = (Qi/Ti) * 60$$

With:

Q: Quantity of the landed or discarded species during haul (i); Ti: Duration of the fishing operation in minutes (i).

For each stratum, an average hourly (Rh) was calculated according to the following formula (Grall & Coïc, 2006):

$$\overline{Rh} = \sum \frac{Rhi}{ni}$$

Specific Richness (SR)

This index was used to analyze the taxonomic structure of the stand and to distinguish spatial and temporal variations. Species richness represents the total (or average) number of species per unit of area (Grall & Coïc, 2006), while SR refers to the number of caught species during the haul.

Shannon Wiener Index (H')

This index measures diversity based on the number of species and the abundance of specimens for each species (Grall & Coïc, 2006). It is calculated according to the following equation:

$$H' = \sum_{i=0}^{S} Pi \log Pi$$

With:

Pi: proportional abundance or percentage of the species Pi = ni/N; S: total number of species; i: number of individuals of a species in the sample; N: total number of individuals of all species in the sample (Grall & Coïc, 2006).

Pielou's Clue (J')

The equitability index measures the distribution of individuals within species. It is calculated according to the following equation:

$$J' = H'/H'max$$

With: $H' \max = \log S$ (total number of species).

This index ranges between 0 and 1. If J' is close to zero, one species is dominant, and if it is close to one, the species is equally distributed (Grall & Coïc, 2006).

Mapping of Fishing Areas Using GIS

Data related to the fishing areas have been analyzed using QGIS software. These data are georeferenced through GIS synthesis. This software was also used for the representation of hourly yields of discards by area. The fishing maps were created using QGIS software (version 3.18) and georeferencing (GPS-Way points).

Results and Discussions

Quantitative Study

In this study, we conducted four surveys in January, February, March, and April 2022. In total, we made 24 benthic trawl hauls, which correspond to 65 hours of fishing. The total landing of commercial species was 1.104kg, which represents an overall hourly yield of 16.98kg/h. During the various trawling operations, the commercial species catch was accompanied by the fishing of a large quantity of non-commercial marine organisms having a low commercial value (discards). The analysis of the 24 samples allowed us to identify 12 commercial species and 27 non-commercial species which were discarded in variable quantities. The quantitative and qualitative composition of these discards were analyzed, and the mapping of the discard yield showed variability between 2kg/h and 9kg/h depending on depth. The maximum (9kg/h) was detected in different areas at different depths (150m-200m). In the north (GSA12), the hourly yield gradually decreases (2kg/h at 200m depth). The discards reported by Ben Jarray et al. (2018) in GSA14 (Gulf of Gabes) account for 53% of the total catch. According to the author, 27% of Tunisia's catch is being discarded. The hourly yield of discards is estimated to be 17.21kg/h in Tunisia's southern region (GSA14), while it is 4.63kg/h in Tunisia's northern region (GSA12).

During the fishing survey which was conducted in the Gulf of Gabes, a production of 1167.2Kg was accompanied by a significant quantity (2602Kg) of small immature fish discarded, associated with a quantity of benthos evaluated at 3155.17 Kg (average hourly yield 49.65kg/h) (Ben Hadj Hamida et al., 2014). These values are higher than in the north of Tunisia (GSA12), which can be attributed to a lack of diversity and abundance of benthic organisms in this region compared to the south (GSA14). According to Ben Hadj Hamida et al. (2014), fishing one kilogram of marketable species resulted in the discard of 2.7 kilograms of small individuals. The discarded species in GSA14 are primarily *D. annularis*, *P. erythrinus*, *P. caeruleostictis*, spicarel and *B. boops*.

In January, a 9.2 kg discard, representing 16% of the total catch per haul, accompanied a 56kg commercial species catch. *S. solea* (7%), *T. trachurus* (18%), and *M. merluccius* (14%) constitute the most discarded species. The amount of commercial fish discarded per haul was 3.56kg, resulting in an hourly yield of 0.6kg/h. Squid (*Loligo vulgaris*) represents cephalopods, with an hourly yield of 0.3kg/h. The discard in January survey includes some benthic species as well. This category of discard has an hourly yield of 0.643kg/h.

During the February survey, haul fishing operations were carried out at various depths (150m, and >200m). 300kg of commercial species were indeed landed, but 92kg of marine organisms were discarded, accounting for 31% of the total catch. The hourly yield of discards in February was 6.13kg/h, which is higher than in January. Fish (3.22kg/h), Selachians (2.72kg/h), and Crustaceans (0.53kg/h) make up the majority of discards (commercial and non-commercial). The highest densities of commercial species discarded were found between 150m and 200m (3kg/h) of depth. In areas where the depth per haul was greater than 200m, the yield dropped to 1.5kg/h.

Discard was minimal (2kg/h) in areas at depths less than 150m. This yield increased to (12kg/h) in areas where the depth was less than 200m.

During the survey that we conducted in March, the landed quantities were 383kg, which represents a yield of 21.27kg/h. The discard was 50kg, corresponding to an hourly yield of 2.77kg/h. For the commercial species, the discard included several groups, such as fish, (19.55kg) with an hourly yield of 1.1kg/h, crustaceans (5.54kg) with an hourly yield of 0.30kg/h, and cephalopods (1.53kg) with an estimated hourly yield of 0.1kg/h. The discard corresponds to the quantity of commercial species and is composed of fish (11.5kg), with an hourly yield of 0.65kg/h, crustaceans (6.5kg) with an hourly yield of 0.36kg/h and selachians (3.9kg), with an hourly yield of 0.220kg/h. The analysis of Figure 15 allows us to affirm that a large quantity of discard is composed of commercial fish (2kg/h). The discard distribution is concentrated in three areas: Zone 1: Estafette (depth 100m - 200m); Zone 2: Esquerquis (mean depth 200m); and Zone 3: Zembra Islands (mean depth 200m). In these areas, the quantities of discard decreased gradually from 2kg/h to 0.5kg/h according to depths ranging from 150m to 200m.

During the surveys conducted in February and March, the most abundant discarded species were fish and crustaceans. Furthermore, it was noted that adverse weather conditions significantly reduced the quantity of discard. Besides, hourly yields differ depending on depth. Indeed, low hourly yields of commercial and discarded species were recorded at depths ranging from 150 to 300m. The highest yield of by-catch species (1.5kg/h) was observed in the northern part of Bizerte at -200m.

A total of 365kg of redfish were landed during the April survey, which represents an hourly yield of 15.86kg/h. A large number of discarded small organisms accompanied the commercial species catch. The quantity of discard is estimated at 152kg (6.60kg/h) and is composed of commercial species, namely fish (102.20kg) with an hourly yield of 4.44kg/h, and cephalopods (2.26kg) with an hourly yield of 0.1kg/h. The highest yields of discards (6kg/h-9.5kg/h) were observed beyond Cani Island, in the north of Bizerte and the Esquerquis area (100m-200m). The hourly yield of discards decreased gradually to reach (3kg/h) in the other areas of GSA12.

The total quantity discarded was 47kg during the April survey, which corresponds to an hourly yield of 2.1kg/h. This discard was made up of two groups of species: fish (0.5kg/h) and Selachians (1.6kg/h). The Esquerquis Area has the highest yield of discarded species (1.5kg/h-3kg/h) (-200m). These yields decreased progressively toward the north (0.5kg/h) at depths less than 200m.

The survey showed that the highest yield of the main groups of discarded species took place in April (15.86kg/h). These values were lower than those obtained from fishing operations in other areas of the Mediterranean Sea. According to a regional study on the importance of discards in the Western Mediterranean, which involved the monitoring of fishing fleets in 7 ports (6 Spanish and 1 Italian), the discarded quantity reached 75.4% and 66.6%, respectively, in the case of larger vessels working in shallow waters (150m depth) in spring and summer (PAS-BIO, 2003). According to a study conducted in Madagascar in 2005, the percentage of fish discarded is high at the start of the fishing season (90% in March-April) (Randriarilala et al., 2005).

Spatial Distribution of the Discarded Commercial Species

Fish

The analysis of discard from the four fishing surveys allowed the generation of geographical maps depicting the spatial distribution of the various commercial groups of species. Most of the discard (3kg/h-7kg/h) consisted of fish, and it was concentrated in two areas (Cane Island and Esquerquis areas) at depths ranging from 150m to 300m.

The highest yields of *M. merluccius* discarded (3kg/h) were observed in the Esquerquis area. This yield can be considered the lowest (0.5kg/h-1kg/h) in the GSA12 Region. *H. dactylopterus*, as depicted in Figure 21, has similar distribution areas to *M. merluccius,* and fishing yields around 1.5 kg/h. Figure 22 depicts the low yields of *Solea solea* in Tunisia's northern region (0.5kg/h).

An increase in bycatch and discards, affecting an important number of species, accompanied the increase in commercial fishery production over the last 50 years. Approximately one-quarter of the commercial marine catch for human consumption would be tossed back into the sea (PAS-BIO, 2003; Pascoe, 1997). This phenomenon of discarding is common throughout the Mediterranean region due to the multi-species nature of its fisheries (FAO, 2004).

Ben Haj Hamida (2004) investigated discard in coastal and offshore fisheries and reported that *M. barbatus* represents the most discarded commercial fish species in Tunisia, particularly in the Gulf of Gabes (GSA14), because of its illegal size. Additionally, Hentati (2012) showed that the discarded quantities accompanying the fishing operations carried out in the southern region of Tunisia (GSA14) are significant and cause damage to the seabed and overexploitation of shallow-water stocks.

Cephalopods

Cephalopods and crustaceans have a wide distribution (more than fish) with a lower yield (0.5kg/h). The small octopus *Eledone cirrhosa* (mean yield 0.4kg/h) is concentrated in the Gulf of Tunis (Dzira El Ouatai) at a depth of less than 150m.

Crustaceans

The golden shrimp *Plesionika martia* has a large area of distribution at depths between 100m and 300m. However, these species present very low yields (0 Kg/h - 0.3 Kg/h).

Spatial Distribution of the Discarded Species

The spatial distribution of species affected by discard is shown in Figures 25, 26, and 27. The most important discarded quantities (2kg/h-3.5kg/h) are fish with high yields in the GSA12 Region. The yields of the discarded Elasmobranchs are also high (1.5kg/h) in the Gulf of Tunis and in the Esquerquis area at depths greater than 100m. On the other hand, the yield of crustacean discards is low (1kg/h) and it occurs in the northern region of Bizerte in areas below 100m depth. The hourly yields of discarded species are higher in-depth areas less than 100m. The distribution of discarded species is determined by environmental conditions, mainly bathymetry and bottom status.

On the other hand, trawling in the Gulf of Gabes (GSA14) often causes discard equal to 50% of the total catches (by-catch, small fish, algae, Posidonia, Cymodocea, etc.) (Jarboui et al., 2005). In this area (GSA14), the average yield per boat had been decreasing in the last 20 years. The number of discarded species is significantly greater than the number of targeted species caught (Bio-div2030, 2021). By-catch includes juveniles of commercially or ecologically important fish species and vulnerable animals; such as sea turtles, sharks, rays, etc. (Bio-div2030, 2021; Jarboui et al., 2020).

The discarding of species that are unfit for consumption (not marketable), or too small, constitutes collateral damage to the ecosystem. This waste of resources represents an additional factor of overexploitation (FAO, 2004).

Benthic trawling in Tunisia's northern region (GSA12) does not cause a significant rate of discard due to the limited availability of aquatic organisms in this area when compared to other regions, such as the Gulf of Gabes (GSA14). Other factors that can influence the amount of discard include the state of the stock, environmental parameters that differ between regions (depth, type of bottom), and the selectivity of the deployed fishing gear. Furthermore, in bad weather, gear maneuvering becomes more delicate, and stability seems to be difficult, resulting in a reduction in discards.

Discard in Mediterranean trawl fisheries varies significantly in quantity and quality, depending on ecological (season, bottom type, and depth), technical (fishing methods and gear), economic (market flooding and species' commercial value), and legal factors (regulations governing fish size) (PAS-BIO, 2003).

Qualitative Evaluation of the Discard

In this section, we analyzed the demographic structures of the main species discarded in the trawling operations carried out in the northern region of Tunisia (GSA12). These species are sole *S. solea*, horse mackerel *T. trachurus*, bottom scorpion fish *H. dactylopterus*, hake *M. merluccius*, scabbardfish *L. caudatus,* and small octopus *E. cirrhossa.*

Solea solea

The demographic structures of *S. solea* discard show that the total length (TL) varies from 8 to 16cm (mean 11.18 ± 182 SD). All the discarded specimens are immature and do not reach the regulatory size (20cm) (El-Aiatt et al., 2019). The most frequent size corresponds to 11 and 12cm which accounts for more than 20%.

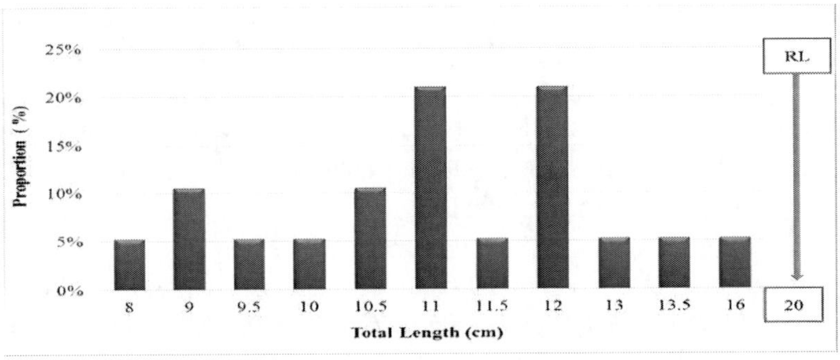

Figure 12. Demographic structure of the discarded *Solea solea* (January - April 2022).

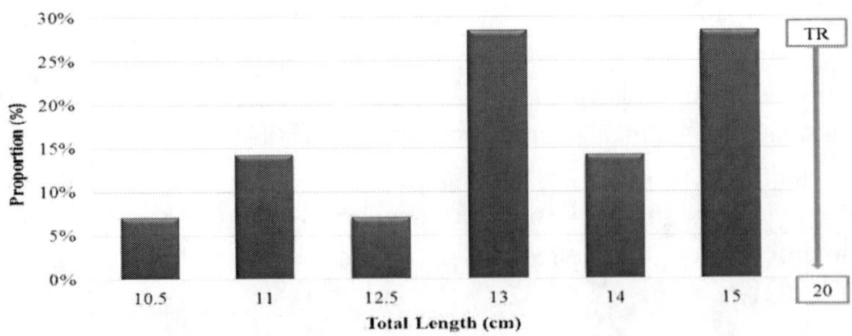

Figure 13. Demographic structure of the discarded *Trachurus trachurus* (January - April 2022).

Trachurus trachurus

Horse mackerel discards are mainly composed of immature individuals with sizes ranging from 10 to 16cm. The L50 estimated by El Achi et al. (2021) ranged from 21.75cm (TL) for males to 22.75cm (TL) for females. According to FishBase (2021), this size is 24.3cm. The most affected sizes by benthic trawling activities in the GSA12 area oscillated between 13 and 15cm (27%). In the long run, the fishing activity using benthic trawls induced overexploitation of the horse mackerel socks in this area.

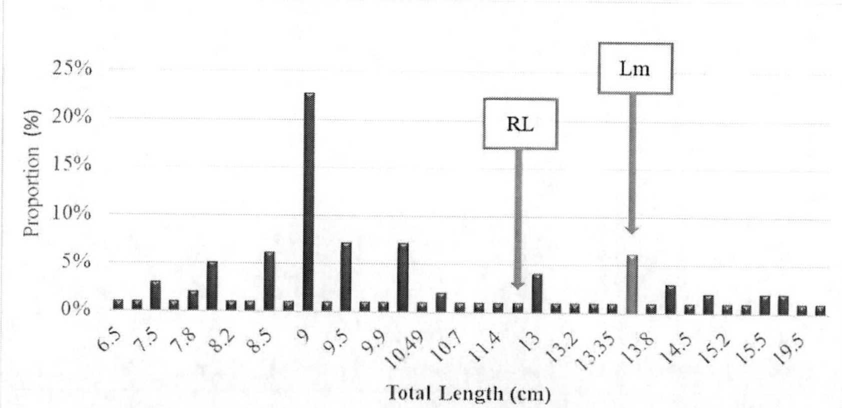

Figure 14. Demographic structure of the discarded *Helicolenus dactylopterus* (January - April 2022).

Helicolenus dactylopterus

The histogram of size frequencies for this species shows that most of the discarded specimens are immature with sizes ranging between 6.5 and 11.5cm. Maturity varies between 32cm for females and 26 for males. According to Nouar (2003) and FishBase (2021) regulatory individual length exceeding 13cm represents 44% of the total catch in the GSA12 Region.

Lepidopus caudatus

The discarded individuals of this species are small. Their demographic structure shows lengths oscillating between 50 and 65cm. The sizes of these specimens are largely lower than the first sexual maturity (100cm) of this species (Estacio et al., 2001). The most represented size classes are 56, 57, and 58cm. The fishing activity of this species seems to be an important parameter inducing overexploitation of the population stock of this species.

Merluccius merluccius

Analysis of the demographic structure of *Merluccius merluccius* shows that the majority of captured specimens' sizes range between 8 and 24cm. The size of these fish is below the first sexual maturity of this species (25cm) (Khoufi, 2015). This observation confirms the negative impact of benthic trawling on this fishery.

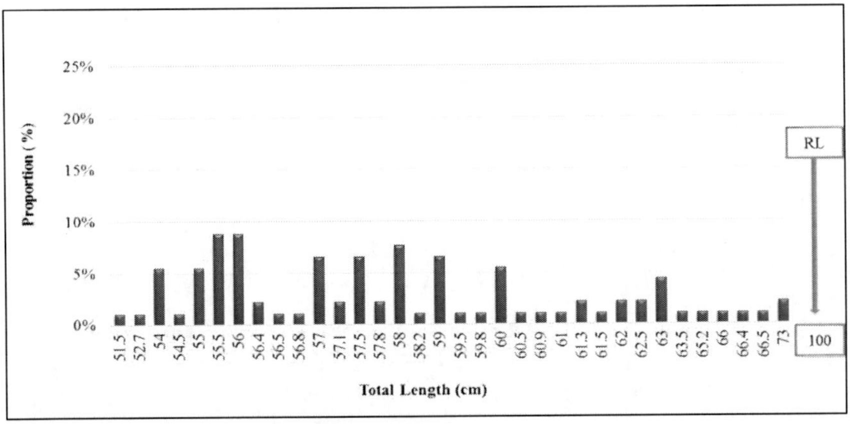

Figure 15. Demographic structure of the discarded *Lepidopus caudatus* (January - April 2022).

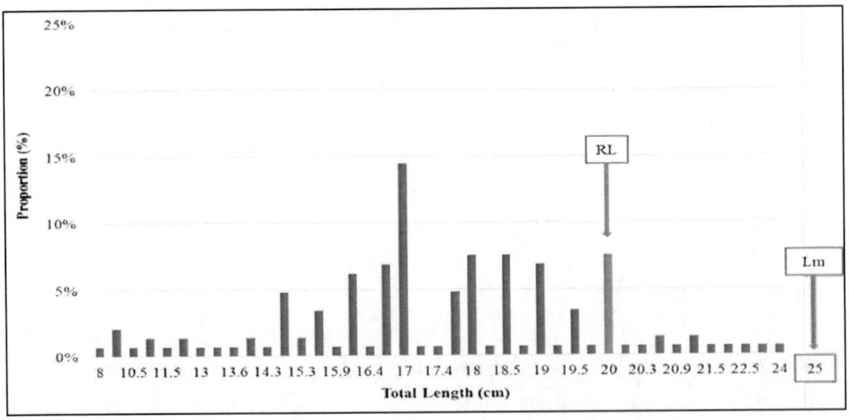

Figure 16. Demographic structure of the discarded *Merluccius merluccius* (January - April 2022).

Eledone cirrhossa

The demographic structure of *Eledone cirrhossa* shows that the mantle length of the specimens captured during the 4 months of the investigation varies from 2 to 26cm. The most frequent size of discard is 2-3cm, which accounts for 28% of the harvest. The rest of the sampled population is composed of individuals that reached the sexual maturity (72%).

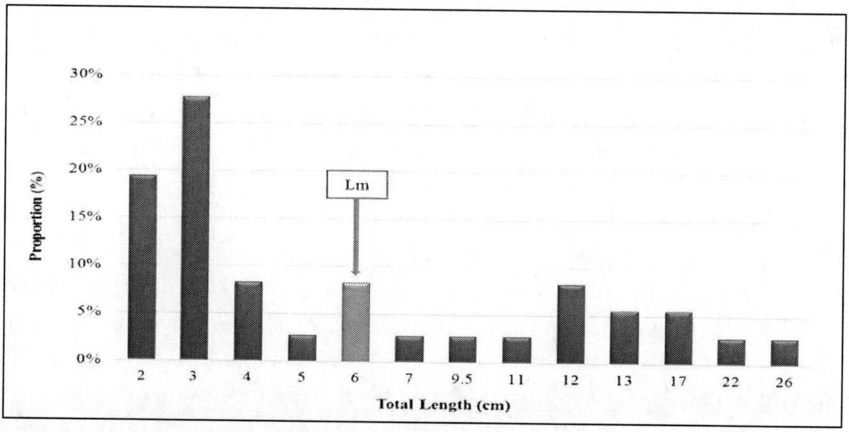

Figure 17. Demographic structure of the discarded *Eledone cirrhossa* (January - April 2022).

Most of the discarded specimens of *E. cirrhossa* are very small, and their lengths are below the first sexual maturity (ICES, 2013). The hydrological and bathymetric parameters in the northern region of Tunisia (GSA12) play an important role in the variation of the distribution of this species.

According to the study made by Ben Hadj Hamida et al. (2014) in the Gulf of Gabes (GSA14), the commercial species were caught with a large number of small-size specimens, which far outnumbered the commercial landings. The majority of the specimens discarded did not reach sexual maturity and thus did not contribute to stock regeneration.

According to Ben Meriem et al. (2005), the assessment of the stocks of the main exploited species in the Gulf of Gabes (GSA14) showed that *M. merluccius*, *M. barbatus*, *M. surmuletus* and *P. erythrinus* are overexploited, and the fishing mortality of the juvenile fraction of the population is particularly high (Bio-div2030, 2021).

Biodiversity in the GSA12 Region

Specific Richness

Species richness is expressed as the number of species per trawl haul. It varied between a minimum of 2 species in March and a maximum of 9 species in April 2022. The trawl hauls that had a high species richness were carried out during March and April surveys. In January, species richness was very poor. This fluctuation in species richness is caused by a variety of factors, including the bathymetry of the fishing area and climatic factors (mainly wind). The hauls with high species richness were made in the Gulf of Tunis, north of Bizerte and Esquerquise, at depths ranging from 100 to 200 meters.

Relative Abundance

The month of January is distinguished by a dearth of species. The *Persisterions cataphractum* had a low percentage (30%). During the February survey, *M. merluccius*, *H. dactylopterus*, and *S. canicula* had a percentage greater than 50%, indicating that these species are rare. *M. merluccius*, *H. dactylopterus*, *Munida rugosa*, and *Pasiphea sivada* were the most abundant species during the March survey (>50%). Species are weakly represented (>30%) in March and they were classified as present. Only *M. merluccius*, *H. dactylopterus*, and *Scyliorhinus canicula* showed a higher percentage of relative abundance (>50%) among the 18 species caught in April trawl hauls. We discovered a new species, *Lepidopus caudatus* (> 49.9%). The remaining species were classified as present (< 29.9%). In addition, we discovered a difference in the haul in each survey ($p < 0.05$).

Shannon-Wiener Diversity Index (H)

The determination of the Shannon-Winner Index allowed for a comparative study of biodiversity in various areas of the GSA12. Based on the analysis made, the Shannon-Winner index varied according to haul. In January, the Shannon-Winner index fell below one, indicating a lack of diversity. The index (H') drops from 0.9 to 0.6 in February 2022. The diversity index H' calculated in these months is similar ($p < 00.5$). H' varies from 0.5 to 0.75 in March. H' fluctuates between 0.6 and 0.9 during the month of April 2022.

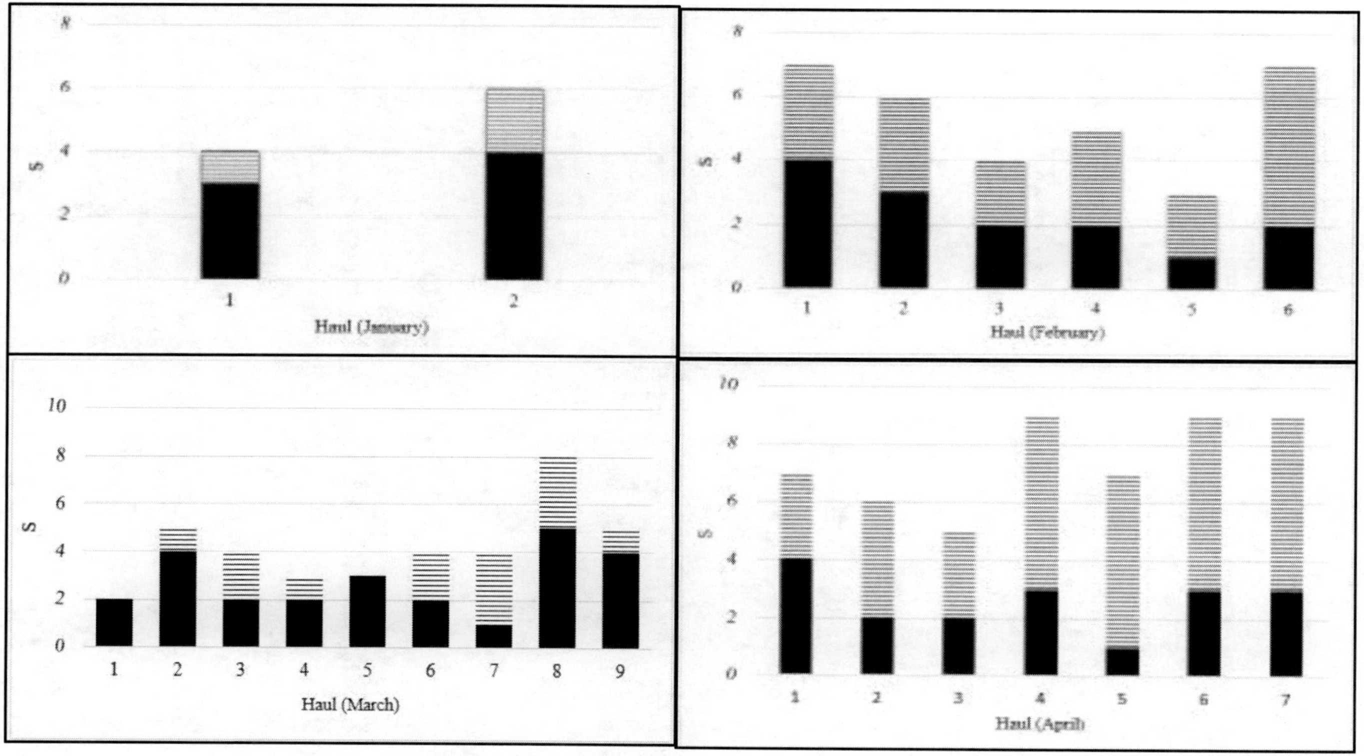

Figure 18. Variability of species richness during January- April 2022 surveys in the GSA12 region.

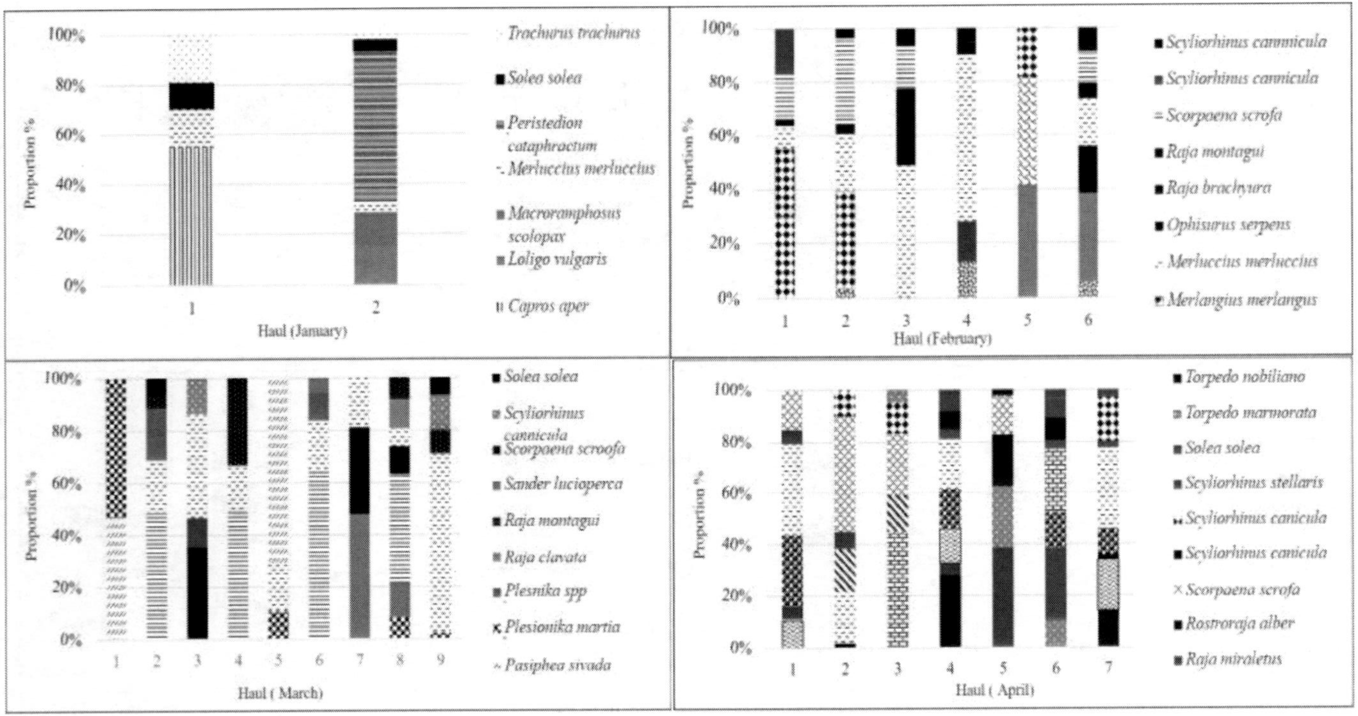

Figure 19. Variability of the relative abundance during January- April 2022 surveys in GSA12 region.

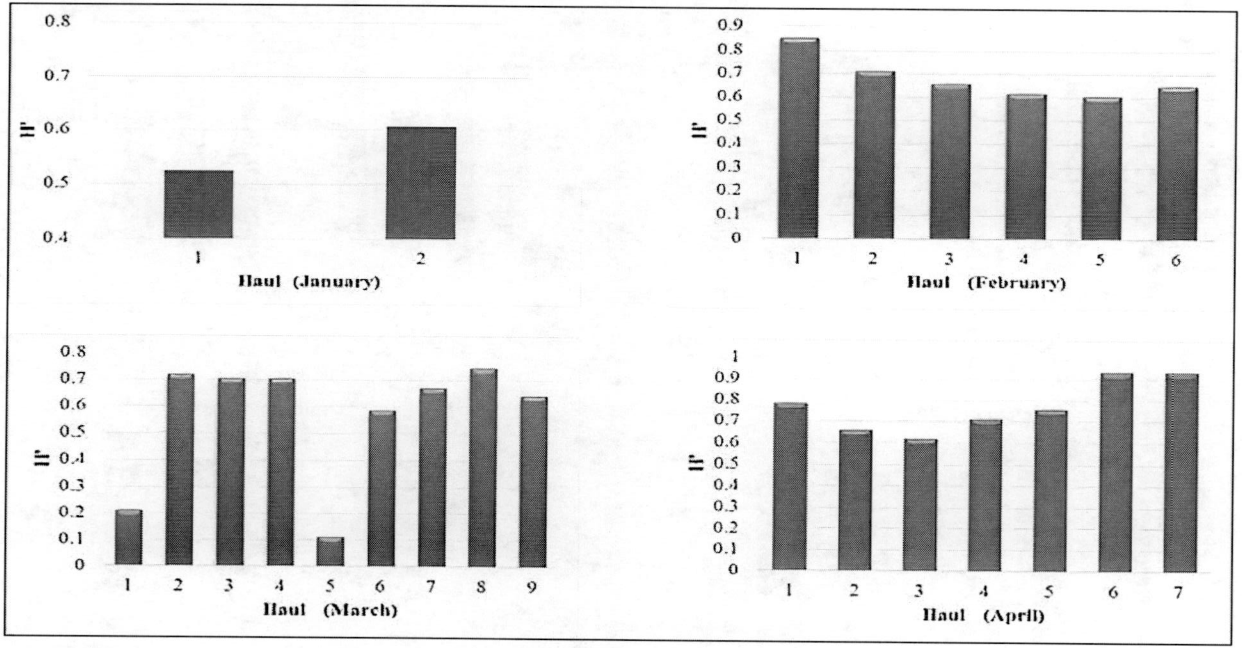

Figure 20. Variability of Shannon-Wiener Index during January- April 2022 surveys in the GSA12 region.

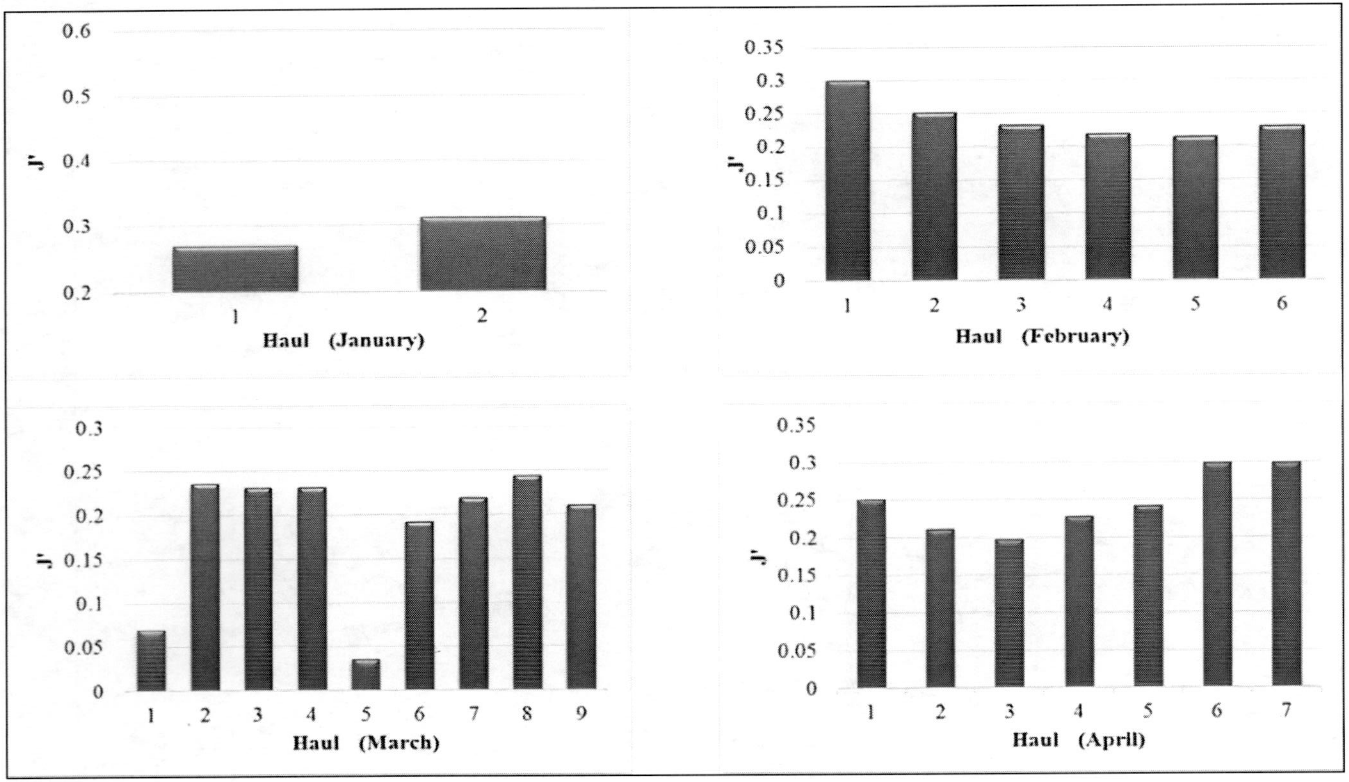

Figure 21. Variability of the Pielou Index during January- April 2022 surveys in the GSA12 region.

In this study, the values of H' are less than one. The decrease in species richness can explain this observation. The ecological status of the fishing areas including Esquerquis, North of Bizerte, and the Gulf of Tunis can be rated as poor according to the study of Simboura and Zenetos (2002). The results show that the diversity in the north of Tunisia (GSA12) is very poor. According to a JICA study (2020) conducted in Tunisia, the by-catch rate by area is estimated at 22% for the north (GSA12), 45% for the center (GSA13), and 33% for the south (GSA14). Additionally, the Gulf of Gabes (GSA14) represents the area with the highest productivity (Jarboui et al., 2020).

Pielou's Regularity Index (J')

We found that Pielou's Index varied between 0.02 and 0.32 during the surveys conducted for this study. These values are mostly below one, indicating that the GSA12 environment is stable.

Conclusion

The current study aims to gain insight into the state of the fishery in Tunisia's northern region (GSA12), by mapping the discards caused by benthic trawlers and indicating the spatial distribution of the different discarded species. To investigate the discard in the GSA12 Region, we conducted a sampling survey on board professional fishing trawlers, which allowed us to create maps of the spatial distribution of hourly discard yields. We also determined the qualitative and quantitative composition of the discarded species. The discard was divided into two categories: "commercial species" and "non-commercial species." The primary findings of this study are listed below:

- Discards represent 21% of the total catch.
- The overall hourly yield of the total discards is 4.63kg/h.
- Fish represent the major part of the discards.
- *M. merluccius* is the commercial species discarded with the highest yield (3kg/h) in the Esquerquis area and at depths greater than 200m.
- Cephalopods and crustaceans have lower yields.
- The yields of the discarded species (3.5kg/h) (Fish, Elasmobranchs, and Crustaceans) are mainly recorded in the Gulf of Tunis and Esquerquis areas at depths greater than 100 m.

- The ecological indices (species richness and abundance) analysis revealed that the number of species varies depending on the period (minimum 2; maximum 9 species).
- The lowest specific diversity (N = 2) was recorded in January, while the highest one (N = 9) was reported during the April fishing survey.
- Shannon Winner's Diversity Index was below 1 during the four fishing surveys. The ecological state in some areas of the northern region of Tunisia (GSA12) can be qualified as poor.

In conclusion, the findings of this study demonstrated that the quantity and quality of the discarded marine organisms vary according to depth, sea state, and fishing area. It is important to note that the amounts of discards recorded during this study in Tunisia's northern region (GSA12) are low due to the type of the deployed fishing gear.

Additionally, fisheries in the GSA12 Region seem to be at their peak of exploitation and require closer monitoring to avoid overexploitation.

Acknowledgments

All our gratitude goes to Mr. Sabri Jaziri and Mr. Okbi Rjeibi assistant professors at the National Institute of Marine Sciences and Technologies (INSTM) for their support. We would also like to thank the owners of the fishing units of the North region of Tunisia. The authors also wish to extend their thanks to the Fishing District of Bizerte. Our gratitude also goes to our colleagues Dhaker Troudi and Hajer Zarrouk for their help.

References

Abdesselem, F. (1998). Determination of coastal fishing effort in the northern region of Tunisia. In: Lleonart J (ed.) Dynamics of marine populations Zaragoza: CIHEAM. *Cahiers Options Méditerranéennes* [*Mediterranean Options Notebooks*], 35, 283-299.

Alpert, P., Krichak, S. O., Krishnamurti, T. N., et al. (1996). The relative role of lateral boundaries, initial conditions and topography in mesoscale simulations of lee cyclogenesis. *Journal of Applied Meteorology*, 35, 1091-1099.

Amrouni, O., Hermassi, T., Adbdeljaouad, S., et al. (2014). Contribution of grain-size trend to sediment of a microtidal beach: Case of the Gulf of Tunis bay (Cape Ferina – Cape Gammarth, Tunisia). Research *Journal of Environmental Sciences*, 8(4), 161-177.

Anonyme. *Assessment of Tunisian fishery resources (1996-2002). Final report: demersal species, pelagic species and socio-economic indicators*. National Research Project, INSTM, Tunisia, 2, 250p; 2002.

Astraldi, M., Gasparini, G. P., Vetrano, A. et al. (2002a). Hydrographic characteristics and interannual variability of water masses in the central Mediterranean: a sensitivity test for long-term changes in the Mediterranean Sea. *Deep-Sea Research*, I (49), 661-680.

Bdioui, M., and Mrabet, R. (2009). Adaptation and testing of a pelagic trawl with very large meshes with four equal faces for a power of 650 HP. *Bull Inst Nat Sci Techn Mer de Salammbo*, 36, 01-11.

Ben Hadj Hamida, O. B. A., Ben Hadj Hamida, N., Ghorbel, M., et al. (2014). *Evaluation des ressources halieutiques dans le golfe de Gabès par chalutage benthique expérimental: campagne Hannibal* (novembre 2013-janvier 2014). Journée Nationale sur la valorisation des résultats de la Recherche dans le domaine de la Pêche et de l'Aquaculture, Sidi Thabet, [*Assessment of fishery resources in the Gulf of Gabès by experimental benthic trawling: Hannibal campaign* (November 2013-January 2014). National Day on the Valorization of Research Results in the Field of Fisheries and Aquaculture, Sidi Thabet] Tunisia, 121p.

Ben Haj Hamida. (2004). *Contribution to the study of the impact of benthic trawling on the biotopes and marine resources of the Gulf of Gabes*. Master's thesis, National Agronomic Institute of Tunisia (INAT), 142p.

Ben Ismail, S., Sammari, C., Béranger, K. et al. (2010). *Atlas of hydrological data of the Tunisian coasts.* National Institute of Marine Sciences and Technologies (INSTM), Tunisia, 22p.

Ben Ismail, S. (2014). *Study of the dynamics in the Central Mediterranean Sea*. PhD thesis, University of Tunis El Manar, Tunisia, 245p.

Ben Jarray, F., Marouani, S., and Jarboui, O. (2018). Adoption of the on-board observer method for monitoring offshore fishing activity in the Gulf of Gabes region: case of the first period of the shrimp fishing campaign (15 MAY - 30 JUNE). *Bull Inst Natn Scien Tech Mer de Salammbô*, (Special Number).

Ben Maiz, N. (1995). *National study on the biological diversity of marine and aquatic flora in Tunisia.* Project MEAT/PNUE/GEF, Ministry of Environment and Territorial Planning, Tunisia, 78p.

Ben Meriem, S., Gharbi, H., and Bedoui, R. (2005). The multispecific fisheries of the Gulf of Gabès: State of exploitation of the main benthic species and management elements. In: *From short to long term: scales of fisheries analysis and management. 7th Fisheries Forum*. Nantes, France, 7p.

Ben Mustapha, K., Hattour, A., Mhetli, M. et al. (1999). Bionomics of the Infra and Circalittoral stages of the Gulf of Gabès. *Bull Inst Natn Scien Tech Mer de Salammbô*, 26, 5-48.

Bio-div2030. (2021). *Study on the identification of drivers of biodiversity loss and key sectors impacting biodiversity in Tunisia.* Provisional Report -V1 Impact of maritime economic sectors on the erosion of Tunisian biodiversity. Draft - Oréade-Brèche, 58p.

Boudouresque, C. F. (1997). *Situation of marine and lagoon biodiversity in Tunisia (Part 2). In: Marine and lagoon biological diversity in Tunisia. State of current knowledge, recommendations for a national conservation and sustainable management strategy*.

Ministry of the Environment and Regional Planning, Tunisia, UNEP (RAC/SPA), 154p.

Bradai, M. N., Quignard, J. P., Bouain, A., et al. (2004). Autochthonous and exotic ichthyofauna of the Tunisian coasts: Census and biogeography. *Cybium*, 28(4), 315-32.

Chemmam-Abdelkader, B., Kraiem, M. M., and Abed, A. (2003). Linear and weight growth in *Dentex gibbosus* (Teleostei, Sparidae) from the Tunisian coasts. *Bull Inst Nat Sci Techn Mer de Salammbo*, 8, 76-79.

Cherif, M. (2014). *Red mullet from the Tunisian North coast: biology, exploitation and population dynamics*. PhD thesis, National Agronomic Institute of Tunisia (INAT), 204p.

DGPA. (2020). *Yearbooks of statistics of fishery products in Tunisia, Directorate General for Fisheries and Aquaculture (DGPA)*. Ministry of Agriculture and Water Resources (M.A.R.H), Tunisia.

El Achi, A., Nafia, M., Manchih, K., et al. (2021). Reproductive biology of horse mackerel *Trachurus trachurus* (Linnaeus, 1758) in the North Atlantic Moroccan coast. *Egyptian Journal of Aquatic Biology & Fisheries*, 25(3), 647-666.

El-Aiatt, A. A. O., Shalloof, K. A. S., and El-Far, A. M. (2019). Reproductive biology of the common sole, *Solea solea* in Southern East Mediterranean, Bardawil Lagoon, Egypt. *Egyptian Journal of Aquatic Biology and Fisheries*, 23(1), 403-411.

Ennouri, R., Zaabou, N., Fertouna-Bellakhal, M., et al. (2015). Assessing trace metal pollution through high spatial resolution of surface sediments along the Tunis Gulf coast (southwestern Mediterranean). *Environmental Science and Pollution Research*, 23(6), 5322-5334.

Estacio, S., Mendonca, A., Krug, H., et al. (2001). Aspects of the reproduction of six exploited demersal fish species in the Azores archipelago. *Arquiptlago. Life and Marine Sciences*, 2(Part B), 83-94.

Euromodel Group. (1995). Progress from 1989 to 1992 in understanding the circulation of the western Mediterranean. *Sea Ocea Acta*, 18(2), 255-271.

FAO. (2004). *Situation mondiale des pêches et de l'aquaculture*. [State of world fisheries and aquaculture.] Sofia, 69p;

FAO. (2023). *Characteristics, structure and resources of the sector*. Text by HAMZA H. [onlíne], version [02/2023], Roma. www.fao.org/fishery/es/countrysector/tn/en, Accessed (02/2023).

Fischer, w., Bauchot, M. L., and Schneider, M. (1987). FAO Species Identification Sheets for Fishery Purposes (Revision 1). Méditerranée et mer Noire. *Zone de pêche*, 37(1), 267-268.

Froese, R., and Pauly, D. (2021). *FishBase*: World Wide Web electronic publication. www.fishbase.org, version (10/2021), Accessed (02/2023).

GIPP. (2019). *Strategic Plan for the Development of Exports of Fishery Products by 2025*. AGRO-Services, Tunisia.

Grall, J., and Coïc, N. (2006). *Synthesis of methods for evaluating the quality of benthos in coastal environments*. European University Institute of the Sea. University of Western Brittany, Laboratory of Marine Environmental Sciences, Brittany, 7-86p.

Guérineau, L., Rochet, M., and Peronnet, I. (2009). *Overview of discards in French fisheries*. Ifremer 49p.

Hentati, Z. (2012). *Quantitative and qualitative assessment of cephalopod trawling discards in the Gulf of Gabès*. Master's thesis, Higher Institute of Fisheries and Aquaculture of Bizerte (ISPAB), Tunisia, 121p.

Herrmann, M. (2007). *Formation and future of water masses in the North-Western Mediterranean: influence on the pelagic planktonic ecosystem, interannual variability and climate change*. PhD thesis, Toulouse III Paul Sabatier University, Aerology Laboratory, France, 308p.

ICES. (2013). *Report of the Working Group on Cephalopod Fisheries and Life History (WGCEPH)*. Scicom steering group on ecosystem functions (SSGEF), Caen, France, 225p.

Jabeur, C. (1999). *La pêche dans le golfe de Gabès : interactions techniques entre les métiers et exploitation partagée du rouget rouge (Mullus surmuletus L., 1758)* [Fishing in the Gulf of Gabès: technical interaction between professions and shared exploitation of red mullet (*Mullus surmuletus L*., 1758)]. PhD thesis, University of Western Brittany, Faculty of Sciences of Brest, France, 161p.

Jarboui, O., Djabou, H., Bernardon, M., et al. (2020). *Implementation of the ecosystem approach for the development of artisanal fisheries on the island of Djerba - Tunisia*. Rome, FAO 95p.

Jarboui, O., Zamouri, N., Ben Hadj Hamida, N., et al. (2005). *Study of benthic trawling discards on the Tunisian coasts*. Laboratory of Living Marine Resources National, Institute of Marine Sciences and Technologies (INSTM), Tunisia, 4p.

JICA. (2020). *Report of the preparatory study for the construction project of surveillance vessels for the management of fisheries resources in the Republic of Tunisia*. Department of Rural Development, Japan International Cooperation Agency 20-003, Tunisia, 242p.

JORT. (1995). *Official Journal of the Republic of Tunisia. Order of the Minister of Agriculture of September 28, 1995, regulating the practice of fishing*. JORT (80) 6 10/95, 1896-1899, Tunisia.

Khoufi, W. (2015). *Biologie, dynamique des populations et gestion des pecheries du merlu (Merluccius merluccius rafinesque, 1810) en Tunisie* [*Biology, population dynamics and management of hake (Merluccius merluccius rafinesque, 1810) fisheries in Tunisia*]. PhD thesis, National Agronomic Institute of Tunisia (INAT), 193p.

Lynne, D., Talley George, L., Pickard William, J. et al. (2011). *Descriptive Physical Oceanography (Sixth Edition)*. Chapter 9 - Atlantic Ocean. Academic press 555p.

M'rabet, R., Bdioui, M., and Ben Naceur, L. (2003). Testing and development of a large vertical opening (GOV) bottom trawl. *Bull Inst Natn Scien Tech Mer de Salammbô*, 30, 15-21.

Mili, S., Ghanem, R., Ennouri, R., et al. S Biological aspects of the Bluntnose Sixgill Shark, *Hexanchus griseus* (Bonnaterre, 1788) in Tunisian waters: implications for fishery management. *Journal of new sciences, Sustainable Livestock Management*, 15(2), 333-345.

Morizur, Y., Pouvreau, S., and Guenole, A. (1997). *Discards in artisanal fishing in the Western Channel.* Study report. Fishery Resources Department IFREMER, France, 127p.

Nouar, A. (2003). *Ecology, biology and exploitation of a species belonging to the family of Scorpaenidae Helicolenus dactylopterus Algerian coasts.* PhD Thesis, Houari-Boumediene University of Science and Technology (USTHB), Algeria, 154p.

PAS-BIO. (2003). *Project for the preparation of a Strategic Action Plan for the Conservation of Biodiversity in the Mediterranean Region: Impacts of fishing practices in the Mediterranean (Impact on sensitive species and habitats, technical solutions and recommendations).* Regional Activity Center for Specially Protected Areas, CAR/ASP, 133p.

Pascoe, S. (1997). *Bycatch management and the economics of discarding.* FAO Fisheries Technical Paper n° 370, FAO.

Quignard, J. P. (1978). Introduction to Mediterranean ichthyology: general aspect of the population. *Bull Off Natn Pêch Tunisie*, 2(1-2), 3-21.

Randriarilala, F., Rafalimanana, T., and Caverivière, A. (2005). *Bycatch of industrial and artisanal shrimpers.* Madagascan coastal prawns, IRD, 262p.

Riedel, W. R. (1959). Siliceous organic remains in pelagic sediments. In: Riedel HA (Editor), Silica in sediments. *Soc Econ Paleontol Mineral, Spec Publ*, 7, 80-91.

Rjeibi, O. (2012). *Biology and population dynamics of the red lobster Palinurus elephas caught on the northern coast of Tunisia.* PhD thesis, National Agronomic Institute of Tunisia (INAT), Tunisia, 281p.

Romdhane, M. S. (1993). Techniques and fishing gear used on the northern coast of Tunisia. Fishing in Tunisia: coastal fishing and the environment. *Cahier du CERES - Série Géographique*, 8, 195-210.

Roussenov, V., Stanev, E., Artale, V., et al. (1995). Seasonal model of the Mediterranean Sea general circulation. *J Geophys Res*, 100(13), 515-13.

Sammari, C., and Brahim, M. (1996). Hydrodynamics of the North-Tunisia/Sicily/Sardinia region during spring 1995. *Bull Inst Natn Tech Océanogr Pêche Salammbô*, 23, 5-32

Sammari, C., and Gana, S. (1995). Review of hydrodynamics off the Tunisian coast. *Bull Inst Natn Tech Océanogr Pêche Salammbô*, 22, 10-31.

Simboura, N., and Zenetos, A. (2002). Benthic indicator to use in ecological quality classification of Mediterranean soft bottoms marine ecosystems, including a new biotic index. *Mediterranean Marine Science*, 3(2), 77-111.

Würtz, M. (2010). *Mediterranean Pelagic Habitat: Oceano graphic and Biological Processes, An Overview.* Gland, Switzerland and Malaga, Spain: IUCN, 90p.

Chapter 5

Parasitic Diseases Encountered in Rainbow Trout (*Oncorhynchus mykiss*)

Mustafa Öz*
and Enes Üstüner
Department of Fisheries and Diseases, Faculty of Veterinary Medicine, Aksaray University, Aksaray, Turkey

Abstract

Fish diseases are one of the most important problems encountered in aquaculture, causing very serious financial losses. Occurrence of deaths and treatment for diseases after the emergence of the disease increases the production costs. Diseases are not seen as a major problem in fish populations in natural environments and sick fish are consumed by another predator. However, in the production systems, stocking the fish very intensely causes the disease to spread more quickly. Rainbow trout, which is grown in many different parts of the world, is one of the fish species that is consumed with great pleasure. Many diseases, including bacterial, viral, parasitic and fungal diseases, are seen in rainbow trout. Parasites are not of great importance in natural conditions. In aquaculture, parasites cause great problems because the fish live in crowded and stressful environments. Parasites seen in fish are generally divided into three groups; protozoa, helminths and arthropods. Parasites can be found on the external surface of the fish as well as in the internal organs. Metazoan parasites are the most numerically seen in rainbow trout. Some parasite species are important for public health due to their zoonotic character. When zoonotic parasites are eaten raw or undercooked, people experience a variety of symptoms. As a result; in

* Corresponding Author's Email: ozmustafa@aksaray.edu.tr.

In: Advances in Marine Biology. Volume 6
Editors: Adam Kovács and Patrik Nagy
ISBN: 979-8-88697-996-1

this book section, a contribution was made to the aquaculture sector by covering the parasitic diseases that are common in rainbow trout, the ways to prevent and control these diseases, the treatment methods of the diseases and the life cycles of the parasites.

Keywords: parasitic diseases, rainbow trout, zoonotic parasites

Introduction

Although the growth rate of the aquaculture sector has decreased for some years, the fishing sector is still ahead of many sectors (Orobets et al. 2019). It is thought that aquaculture production will continue to grow with the increase in consumption with the increasing population (Bostock et al. 2010). Salmonids (most *Salmo salar*) producers (excluding Asia) are Norway and Chile, 2% of global production (D'Agaro, Gibertoni, and Esposito 2022). Global trout production was 940 thousand tons in 2019 (FAO 2022). The most grown species is rainbow trout, accounting for 97% of the total volume (D'Agaro, Gibertoni, and Esposito 2022). One of the biggest obstacles to the increase in this production is diseases. Parasitic diseases are rarely pathogenic due to the large area of the natural aquatic environment and the widespread of the fish population. In fish farms, this situation is the opposite, causing great economic losses. There are multiple factors affecting the health of fish, some of which are; physiological condition, host, cultivation, pathogen, aquatic environment and feed content (Subramani and Michael 2017). These factors are shown in Figure 1 below. In addition to knowing the factors that cause parasitic diseases, knowing their life cycle is important in understanding the pathobiology of the disease (Jeney 2017). Some parasites have direct life cycles, while others use aquatic invertebrates. Fish parasites are unicellular (protozoa) or multicellular (metazoan) organisms. Those that settle on the skin, gills and fins are ectoparasites, parasites that settle in internal organs and tissues are found as internal parasites (Yarsan 2020). Some fish parasites are of zoonotic importance. For example, fish-borne zoonotic trematode infections affect more than 18 million people all over the world, especially in Asian countries (Pekmezci, Yardımcı, and Onuk 2017). In this book chapter, information will be given about the most common protozoa and metazoan parasites in rainbow trout.

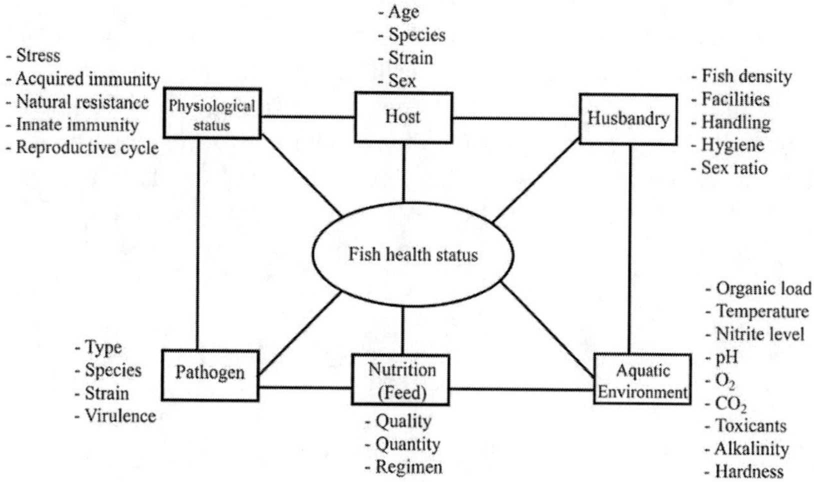

Figure 1. Factors affecting the health status of fish (Kum and Sekkin 2011).

Protozoan Diseases

Ichthyophthiriosis

The causative agent of the disease is Ichthyophthirius *multifiliis*. It is a ciliate species and an important pathogen of fish living in both temperate and tropical regions worldwide (Matthews 2005). Ichthyophthirius *multifiliis* is one of those with high pathogenicity in both freshwater and saltwater fish. Almost all freshwater fish are susceptible to the disease and mortality rate can be as high as 100% in scaleless fish such as catfish (Meyer and Hoffman 1974). The life cycle leaves the host following the growth and development of the trophont on the fish. Then, as a result of contact with water, structures called tomonts and daughter cells called tomites are produced inside the tomonts. With the realization of a certain number of divisions, each tomite turns into structures called theront. These theront structures are then contagious to the host (Matthews 2005). The white points that give the disease its name are due to the feeding of trophonts attached to the skin of the fish from the epidermal cells (Buchmann 2020). Severe infections are lethal in fish, but fish that have survived the disease are immune to the infection (Buschkiel 1910). It is a temperate water disease and the emergence temperature of the disease is 15 – 25°C. The life cycle of the parasite lasts about 3-6 days at 25°C, about 10 days at 15°C, and more than about a month at 10°C (Noga 2010). The circle of life is shown in Figure 2. Salmonids occur in summer at 19°C, when heat stress begins in epidemic fish; but it can go down to 4° in salmonids and other cold water fish (Noga 2010). Lesions up to 0.5 mm in diameter are seen due to epithelial damage from trophonts attached to the fish. It results in suffocation and death in fish due to localization to the gills and damage to the gill filaments (Coyne et al. 2011).

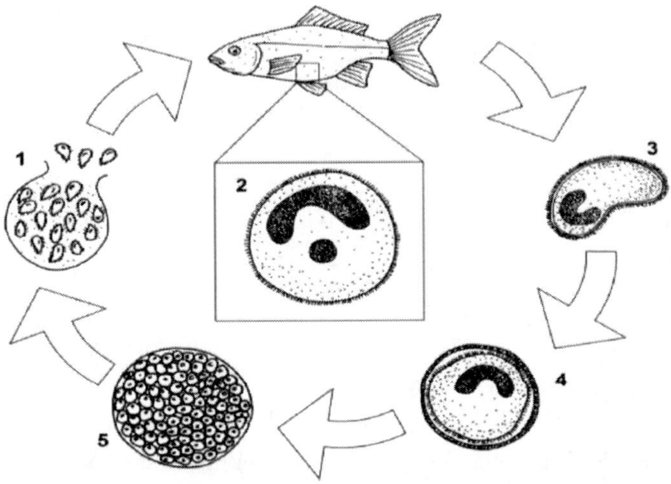

Figure 2. Life-cycle of *Ichthyophthirius multifiliis*. 1) Infective theronts 2) Parasitic trophon stage on fish 3) Exiting tomont 4) Cyst. 5) Dividing tomites within cyst. (Verner-Jeffreys and Taylor 2015).

Epithelial damage and ulceration caused by the parasite may lead to the establishment of secondary infections (Noga 2010). The diagnosis is easily confirmed by microscopic examination of the skin and gills. A few white grubs are removed from the infected fish and then covered with a few drops of water and a microscope slide. The mature parasite has a large, dark horseshoe-shaped macronucleus visible under 100x magnification (Francis-Floyd and Reed 1991). Its microscopic view is given in Figure 3.

Various chemicals, drugs, and natural extracts are used in the treatment of the disease (Picon-Camacho et al. 2012). Formaldehyde is one of the most used chemicals. Formaldehyde affects both fish and aquatic forms of the parasite (Heinecke and Buchmann 2009). It is toxic to both fish and humans when not used in the appropriate dose and time. In a study conducted on *Oncorhynchus mykiss*, it was reported that it thinned the mucus layer and reduced mucus production in fish exposed to 200-300 ppm concentration, making it more susceptible to secondary infections in water (Buchmann, Bresciani, and Jappe 2004). It was reported that for every 5 ppm of formaldehyde used in water, it reduces the dissolved oxygen in water by 1 ppm (Cross 1972). The other most commonly used chemical is sodium chloride. It was reported that the use of 2.5 grams per meter reduces the number of tomont and theront (Lahnsteiner and Weismann 2007). Besides, copper sulphate, potassium permanganate, chloramine-T, hydrogen peroxide, metronidazole, toltrazuril, and various natural extracts are used (Picon-Camacho et al. 2012).

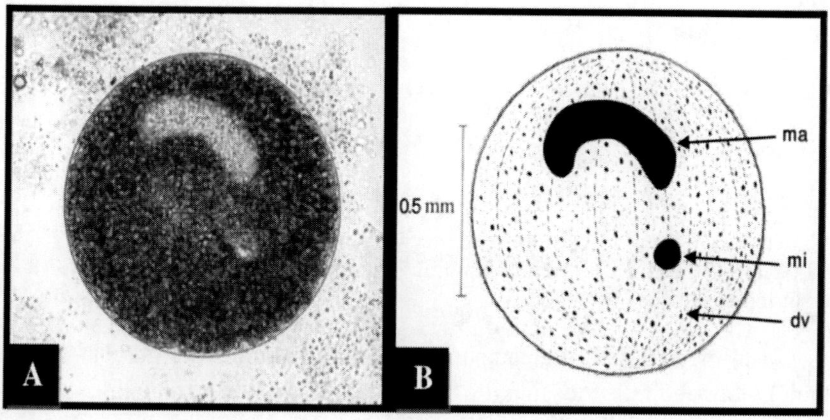

Figure 3. *Ichthyophthirius multifiliis* A) Photomicrograph (50x) B) *Camera lucida* drawing dv= digested vacuole; ma= macronucleus; mi= micronucleus (Abdullah 2013).

Ichthyobodosis

The causative agent of the disease is *Ichthyobodo necator*. It is a type of parasite that can infect freshwater farm fish and freshwater wild fish (Bruno, Noguera, and Poppe 2013). The parasites attached to the fish attach to the epithelial tissue and damage the tissues (Joyon and Lom 1969). After hyperplasia in Malpighian cells and discharge of goblet cells, epidermal shedding occurs (Bullock and Robertson 1982). *Ichthyobodo necator* causes infection in a wide temperature range such as 2-30°C (Noga 2010). It causes infection in fish living in warm waters at low temperatures. It is reported that it dies if the temperature is above 30°C (Noga 2010). Parasite salmonid was noted to cause infection in hatcheries at temperatures between 2°C and 14°C (Awakura, Kojima, and Tanaka 1984). Living in the free and parasitic period, *Ichthyobodo necator* reproduces by dividing into two. Cyst development occurs in both free-living and parasitic forms. Free-living parasites come into contact with the fish with their flagella and a parasitic form develops in about a few minutes (Yarsan 2020). The morphological image of the parasite is given in Figure 4.

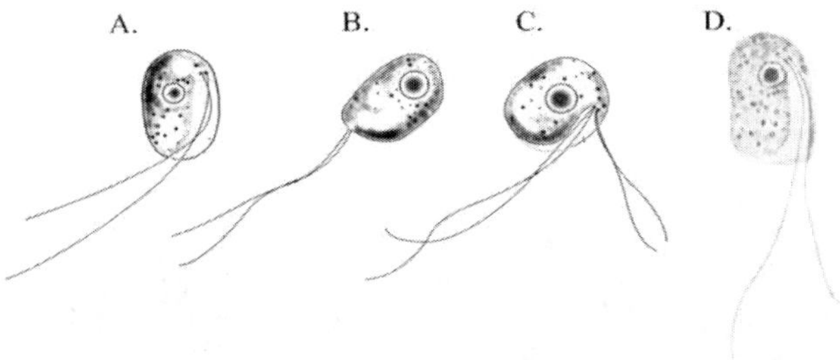

Figure 4. Free forms of *Costia necatrix* sensu Andai (1933). A: Ventral view. Bi-flagellated. B: Dorsal view. C: Ventral view, four flagellated cell. D: Original sketch (Isaksen 2013).

Fish infected with the parasite develop a light bluish or greyish layer on their bodies and fins, thus making the inside of the tank appear to glow. Goblet cells are lost in the epidermis, whereas in salmonids, hyperplasia is seen in Malpighian cells in severe body infection. Necrosis of epithelial cells and collapse of blood vessels can be seen (Woo and Leatherland 2006). In the gills, depletion of goblet cells, hyperplasia, and fusion and shedding of adjacent lamellae can be seen. During healing, numerous eosinophilic granules can be seen on the lamellae (Bruno, Noguera, and Poppe 2013).

Diagnosis is made by seeing the factors in scraping preparations prepared from the skin and gills. The parasite is sensitive and may be damaged during fixation and staining. It is recommended that the preparations are fixed with Shaudinn fixative and passed through ethyl alcohol and stained with Carmin (Becker 1977). 166 ppm formalin for 1 hour or a 15-30 minute formalin bath (1:4000) can be used to eliminate parasites in juvenile salmon fish (Skrudland 1987). 1-2% salt solution can be applied for 20 minutes in trout and carp fry. Raising the water temperature to 30°C can kill the parasites, but it should be considered whether the fish can tolerate this temperature (Erer 2002).

Hexamitosis

Infections caused by Hexamita and Spironucleus genera are considered as hexamitosis (Yarsan 2020). *Hexamita salmonis* and Hexamita intestinalis are important flagellated protozoa of trout (Erer 2002). The parasite is usually seen in the digestive system, especially in the intestines, although rarely in internal organs and tissues (Woo and Leatherland 2006). The parasite has a direct life cycle (Hausmann and Gleeson 2022). The infection is thought to be caused by the cystic form of the parasite or by oral ingestion of trophozoites (Poynton and Morrıson 1990). The trophozoite form of the parasite usually causes lesions in the intestinal lumen (Kent et al. 1992). Changes in diet, low oxygen content in the water and keeping fish of different lengths together have been reported as the reason for increasing the amount of parasites (Vickerman 1990). It is also called hole-in-the-head disease due to the lesion it creates in the head region. The parasite is found in the intestinal flora of healthy fish, and the disease begins to manifest itself when a negative change in health and quality of life occurs in the fish. In the intestines, haemorrhagic enteritis occurs before the simple enteritis picture, and this is characteristic (Yarsan 2020). Hexamita infections, which usually infect the intestines, become chronic since there are few parasites, and clinical findings occur in acute situations where the parasite density increases (Woo and Leatherland 2006). In acute infections, loss of appetite, weakness, lethargy, darkening in colour, and fish swimming towards the water outlet are seen (Balta and Dengiz Balta 2018). In severe infections, samples should be taken from freshly dead fish (from the gastrointestinal tract, internal organs and blood) and dried in air, fixed in alcohol, and a smear should be prepared (Woo and Leatherland 2006). Since the parasite moves very quickly, the addition of viscous protoslo or methyl cellulose can be used to slow the parasite's movements to reduce the parasite's movements (Lee et al. 1985). In treatment, metronidazole is applied as a dip, the common dose is 5-7 mg/L every 24 hours or 25mg/L every 48 hours for 6 hours. Apart from these, nitroimidazoles (benznidazole, ronidazole, triclabendazole, albendazole, mebendazole) or magnesium sulphate can be used orally (Hausmann and Gleeson 2022).

Amoebic Gill Disease

Amoebic gill disease (AGD) of marine-farmed salmonids was first described in Tasmania, Australia (Munday 1986). It caused more than 80% mortality in the 2006 epidemic in Norway due to deficiencies in treatment (Steinum et al. 2008). The causative agent of the disease, *Neoparamoeba pemaquidensis*, has been accepted as the etiological agent for many years (Munday, Zilberg, and Findlay 2001). Numerous experimental studies using *N. pemaquidensis* have been unsuccessful (Morrison et al. 2005). As a result of phylogenetic studies based on 18S rRNA and 28S rRNA gene sequences, the causative agent of the disease was identified as *N. perurans* (Young, Dyková, Snekvik, et al. 2008). The gills extracted from infected fish are approximately 15-40 μm in diameter in a rounded form (Rodger and McArdle 1996). The parasite nucleus stained with Feulgen is approximately 5 μm in diameter and has more than one parasome, and its size is approximately 4 μm (Woo and Leatherland 2006). The circle of life multiplies by dividing into two. Pathogens in water can survive for up to 2 weeks. Pathogens are concentrated in the materials in the cage system. Pathogens remain viable for up to 3 days in dead fish (Yarsan 2020). Infection most commonly occurs between 10 – 20°C of water temperature (Roberts 2012). Infected fish, due to their placement in the gills, are seen in movements to the water surface due to respiratory difficulties (Rodger and McArdle 1996). Proliferation and fusion in the lamellar epithelium and the decrease in the surface area in the gills and the increase in the diffusion distance between water and blood are the main causes of respiratory distress (Marcos-López and Rodger 2020). In chronic and severe cases, thickening of the stratum layer of the ventricle can be observed in the liver (Rodger 2014) and heart due to hypoxia (Powell, Nowak, and Adams 2002). Scoring can be used to determine the severity of infection and treatment method in the diagnosis of the disease (Taylor et al. 2009), but since scoring is open to misinterpretation, it gives inaccurate results in diagnosis (Rozas-Serri 2019). Histology can be used to diagnose the disease, to identify the causative agent, and also to investigate the immune response (Adams and Nowak 2004).

PCR is one of the important diagnostic methods to be used in agent identification. The specific 18S rRNA gene for the detection of N. perurans in gill samples is used for amplification and identification due to its high sensitivity (Bridle et al. 2010; Rozas-Serri 2019). An important advantage of the PCR method over histopathology is that the agent can be found as positive in swap samples taken from fish without histopathological findings (Rozas-Serri 2019; Young, Dyková, Nowak, et al. 2008). In treatment, 800-1300 ppm hydrogen peroxide bath application for 12-20 minutes is an effective treatment method against AGD (Rodger 2014). It should be used with caution as it may cause irritation to the gills (Kiemer and Black 1997). Experimental vaccination studies have been done to stimulate the immune system (Valdenegro-Vega et al. 2015), but AGD has not been very effective in the survival of infected fish (Oldham, Rodger, and Nowak 2016).

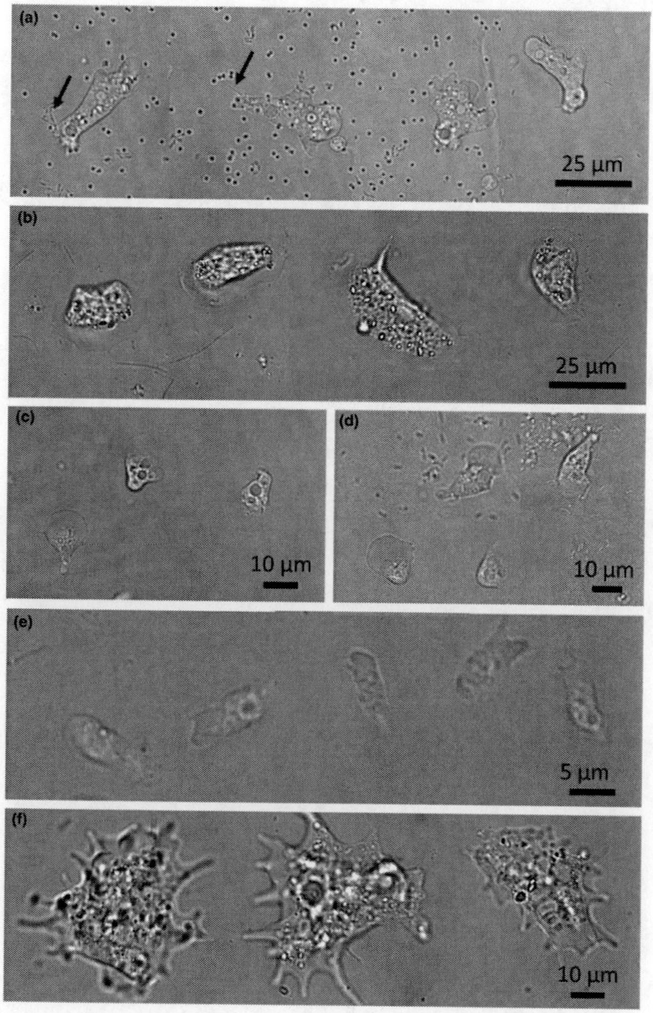

Figure 5. Light microscopy of cultured amoeba strains isolated from gills of rainbow trout with signs of NGD (a) *Naegleria* sp., flagellae (arrows); (b) *Cochliopodium* sp.; (c) *Vannella* sp.; (d) *Ripella* sp.; (e) *Mycamoeba* sp.; (f) unidentified amoeba (Vannetti et al., 2023)

Freshwater Velvet Disease

Piscinoodinium is a disease-causing genus in freshwater fish, and *Amyloodinium* species is an infection in saltwater fish (Yarsan 2020). It has been reported in North America (*P. limneticum),* European *(P. pilulare*) aquarium fish, as well as commercial fish in Asia (Lom and Schubert 1983; Shaharom-Harrison et al. 1990) and South America (Carneiro, Martins, and Urbinati 2002). The life cycle of the parasite consists of 3 stages. Immobile trophonts feed by attaching to the gills and skin of the fish. Trophonts separate from the host to the tomont stage and become cysts. Tomonts produce free-floating dinospores. It is most contagious in the first 24 hours and remains contagious for 6-15 days (Noga 2010). It mostly infects the skin and gills of fish. It is very pathogenic in young fish and can live for months in older fish (Woo and Leatherland 2006). In severe infections, a yellow and rusty colour on the skin is characteristic for the disease (Shaharom-Harrison et al. 1990). Rhizoid structures in trophonts attach to fish epithelial cells and damage them. It worsens with the effect of hydrodynamic force of water on the adhesion surfaces. Cellular damage is thought to attract parasites to the skin, since the trophonts cluster on the damaged skin surface (Woo and Leatherland 2006). In the histopathology of the gills, separation of the epithelium of the gills, degeneration and necrosis of all gill filaments can be seen (Noga 2010). The microscopic image of the parasite is shown in Figure 6.

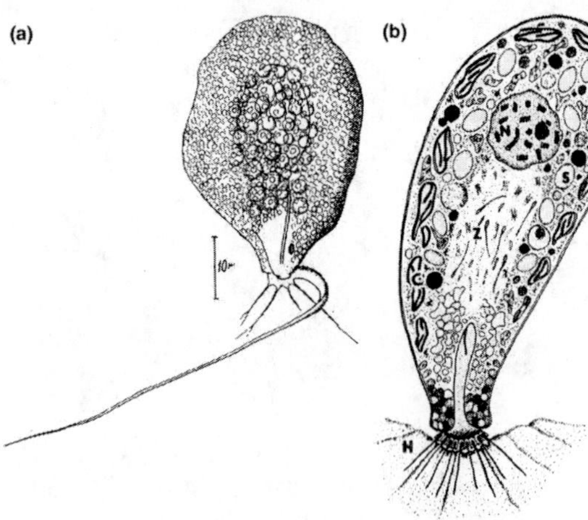

Figure 6. Trophonts of (a) *Amyloodinium ocellatum*, (b) *Piscinoodinium pillulare* (Lieke et al., 2019)

Lesions and colour change in the gill and skin give an idea in the diagnosis. In acute infections, fish may die before skin lesions develop. Definitive diagnosis is the presence of factors in scraping or biopsy material or PCR method is used (Yarsan 2020). It is recommended to increase the temperature by 24-27°C in order to move the fish to another environment, leave the aquarium without fish for 2 weeks and speed up the life cycle (Noga 2010). Dinospores are affected by chemicals added to the water, but the same situation is not effective on trophont and tomont forms, and they are resistant to many chemicals. The most commonly used chemicals are salt, copper compounds, and chloroquine (Yarsan 2020).

Helminth Diseases

Gyrodactylosis

The causative agent of disease (usually salmon, trout) is mostly located on the skin, gills, less often on the fins and nostrils (Yarsan 2020). They are highly host-specific parasites and the species number is estimated at over 20,000 (Bakke, Harris, and Cable 2002). A few species are highly pathogenic to salmonids. In Europe, the most pathogenic species for fish such as Atlantic salmon (*Salmo salar*), brown trout (*Salmo trutta*) and rainbow trout (*Oncorhynchus mykiss*) are *Gyrodactylus salaris* and *Gyrodactylus derjavinoides* (Woo and Buchmann 2012). In this section, these species will be explained in detail. *G. salaris* is a problem in Europe, especially for Norway, and has been reported in 23 European countries (Paladini et al. 2021). The front end of the parasite is two-lobed and has no eye spots. In the posterior adhesion organelle (opishaptor), there are 1 pair of large anchors in the medial, 8 pairs of small anchors on the marginal, and connecting rods in the dorsal and ventral parts (Yarsan 2020). The morphological image of *Gyrodactylus salaris* is shown in Figure 7.

They have a direct life cycle. Gyrodactylids are viviparous and a parasite can live for up to three generations. Transmission from parasites from live or dead fish is more effective than free-living parasites. Transmission of *G. salaris* is due to transportation and stockpiling of parasite-infected fish rather than contaminated water. It completes its life cycle fast. Its survival generally is 1 day and in *G. salaris* it ranges between one to four days (Hausmann and Gleeson 2022). It is usually located on the skin and gills. It causes skin irritation during feeding, which causes turbidity on the skin, as well as excessive mucus production, epithelial hyperplasia, and focal redness due to bleeding in the skin (Kabata 1985). Infection severity is proportional to the number of parasites. The death toll is higher in juvenile fish (Busch, Dalsgaard, and Buchmann 2003). Due to the lesions it creates on the skin, it facilitates the transmission of secondary infections (Bruno, Noguera, and Poppe 2013). In the diagnosis of gyrodactylids, it is recommended that the agent be taken from the skin, gills and oral cavity and fixed for morphological and molecular studies after 95% ethyl alcohol is fixed. Morphologically, adults are 0.3-1 mm long, no eye spots, opisthaptor has 4 large hooks, 8 pairs of side hooks, and the head shape is letter V. Diagnosis of *G. salaris* is usually made with TaqMan PCR, but ISH and RFLP tests can also be used. *G. salaris* must be reported to the OIE when diagnosed (Hausmann and Gleeson 2022).

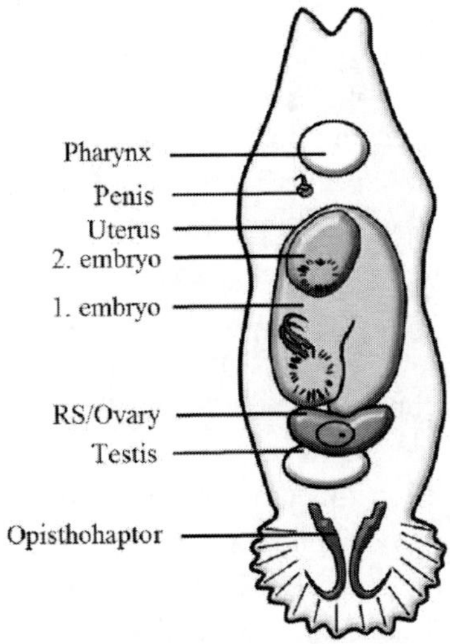

Figure 7. Anatomy of *Gyrodactylus salaris* in a schematic based on a light-micrograph (interference contrast) (Fromm 2014).

Sanguinicolosis

Adult parasites are found in capillaries in the circulatory system and gills of various final host fish (especially carp and trout). The most well-known species are *Sanguinicola inermis, S. armata and S. intermedia* (Yarsan 2020). In North America, *Sanguinicola* spp. has caused severe gill damage in rainbow trout and high mortality in brook trout hatcheries (Roberts 2012). The life cycle is indirect, with fish being the final host. The adult parasite lays its eggs in the blood, and the miracidum, which develops in 1 week in the eggs that go to the gill capillaries with the blood flow, leaves the egg, and passes into the water through the gill capillary epithelium. Miracidium enters water slugs as intermediate hosts and in about 4-5 weeks the infective form for fish, furkocercariae develops. Furkocerca leaves the intermediate host and clings to the gills of the fish in the water. It enters the circulatory system and matures in blood vessels in about 1 month (Yarsan 2020). The morphological image of *Sanguinicola platyrhynchi* is shown in Figure 8.

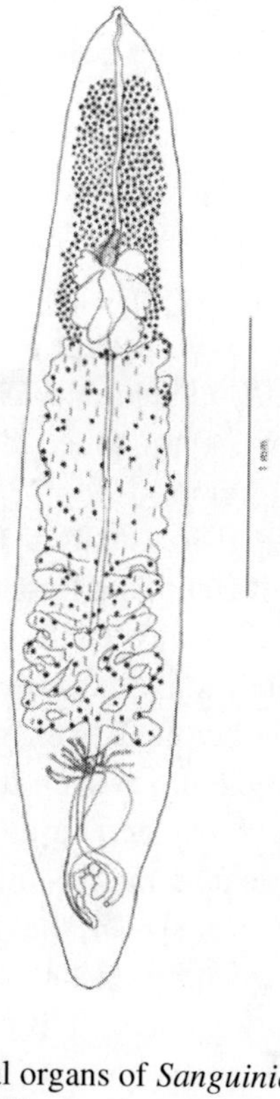

Figure 8. Ventral view of internal organs of *Sanguinicola platyrhynchi* (Guidelli, Isaac, & Pavanelli, 2002)

Sanguinicola spp. is localized in the ventral aorta and branchial arteries of rainbow trout and brook trout. The eggs reach the gill capillaries with the blood circulation and cause ruptures and haemorrhages on the vessel wall where they settle (Bruno, Noguera, and Poppe 2013). Miracidium in the gills can cause mechanical damage, haemorrhage, and necrosis in the heart and kidneys (Bruno, Noguera, and Poppe 2013). Miracidium in the gills is one of the main causes of death in infected juvenile trout (Evans 1974). In chronic infections, adult parasites enter the heart, kidney, and caudal veins through the blood. It is encapsulated here and may cause focal granulomas (Woo and Leatherland 2006). Histopathological changes are most common in juvenile fish and the extent of damage is related to parasite density (Woo and Buchmann 2012). Eggs induce hyperplasia of primary and secondary lamellae by reducing the respiratory surface. It causes vascular obstruction and destruction of veins, ischemia, and necrosis of gills (Woo and Leatherland 2006). While praziquantel is effective, combating intermediate host water slugs is the only proven control (Noga 2010).

Diplostomatosis

It is known as eye worm disease of fish. The disease is seen in salmon, trout, and carp. The most well-known species in the disease are *Diplostomum spathaceum, D. baeri, D. indistinctum* and *D. complanatum* (Yarsan 2020). *Diplostomum* species are 220-300 µm in length and 220-280 µm in width and have an oval appearance (Akıncı 1999). *Diplostomum spathaceum* is a seasonal disease of freshwater fish, usually in the wild and farm-raised. It is a problem with rainbow trout raised in soil ponds and shallow waters (Bruno, Noguera, and Poppe 2013). The life cycle of *Diplostomum* spp. generally consists of 4 parts. The first portion of parasites matures in the gut of the final host. In the second part, they lay their eggs in the aquatic environment with feces. In the third part, miracidium, which develops in eggs excreted with feces, enters the first intermediate host, water slugs (Karvonen and Marcogliese 2020). When it enters the first intermediate host, sporocyst and finally many cercariae are formed in the intestines. In the fourth part, furkocercariae leaving the first host enter through the lateral line of the fish and migrate body. The metacercariae form is mostly located in the eye area of fish and without forming a cyst (Yarsan 2020). Morphological image of *Diplostomum spathaceum* is shown in Figure 9.

The sign of infection is cataract caused by metacercariae in the lens of the eye. If there are too many metacercariae, the damage increases even more and can form the chronic stage known as diplostomiasis. In the case of intense infection, the fish loses their vision, the lens capsule may rupture, or the lens may displace (Karvonen and Marcogliese 2020). The degree of cataract is proportional to the number of parasites in the lens (Karvonen, Seppälä, and Valtonen 2004). Infected fish are less adaptable and swim closer to the water surface.

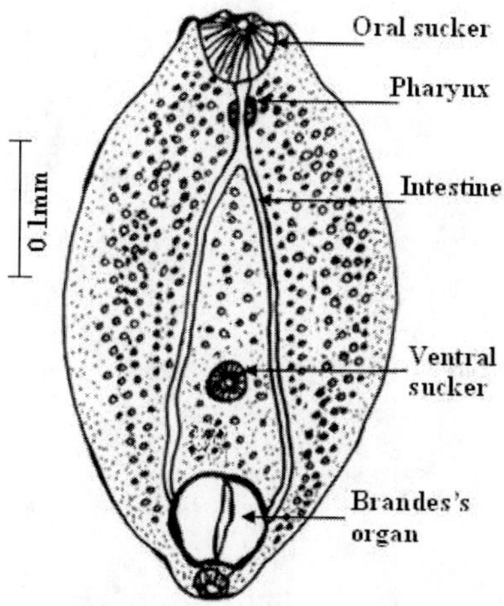

Figure 9. *Diplostomum spathaceum*. Camera lucida drawing. (Shwani and Abdullah 2010).

Blind fish begin to weaken because they cannot take the bait, and they easily become food for waterfowl due to their swimming disorder (Crowden and Broom 1980; Erer 2002; Karvonen and Marcogliese 2020; Seppälä, Karvonen, and Valtonen 2008; Yarsan 2020). In the diagnosis of the disease, clinical observations, fresh vitreous and humor aquosus smear preparations or metacercariae are sought in the histological sections of the eye (Bruno, Noguera, and Poppe 2013; Yarsan 2020).

Clinostomosis

The genus *Clinostomum* (Digenea, Clinostomidae) are digenetic trematodes found in freshwater fish (Calhoun et al. 2020). It forms cysts in the muscles, gills and other organs of fish, and is known as yellow grubs because of these cysts (Bullard and Overstreet 2008). Adult forms of the parasite inhabit the mouth, pharynx, and oesophagus of fish-feeding waterfowl, reptiles, and humans (Simsek et al. 2018). There is a tongue-shaped, flattened, non-segmented mouth and abdominal puller in the form of metacercariae that causes infection in fish. In the cystic form, it is yellowish in colour, elliptical in shape, and approximately 3-5 mm in diameter (Yarsan 2020). The morphology of the parasite is shown in Figure 10.

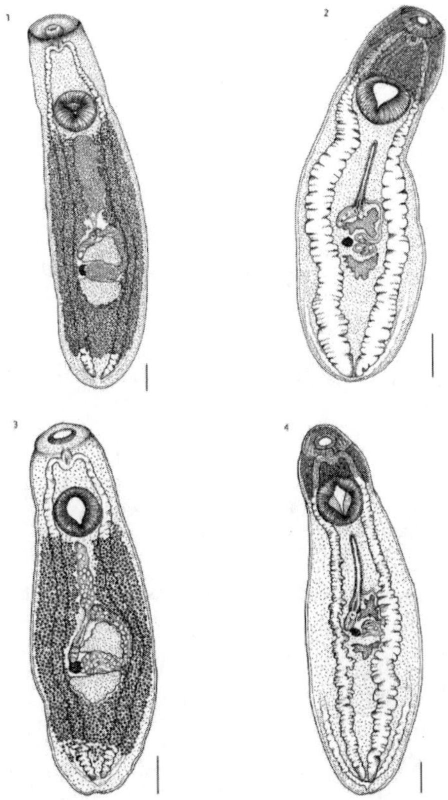

Figure 10. *Clinostomum marginatum*. (1) Adult. (2) Metacercaria, *Clinostomum complanatum*. (3) Adult. (4) Metacercaria. Scale bars = 500 μm (Gustinelli et al., 2010)

Mature forms of the parasite in its life cycle are fish-eating waterfowl. Eggs are released from the final host into the aquatic environment, where they transform into miracidium, then migrate to water snails, where they develop into sporotcysts (Smyth and Smyth 1980). Sporocysts then transform into free-swimming cercariae and attach to freshwater fish and amphibians (McAllister 1990). When they reach the fish, they transform into the metacerca form and become cysts (Kanev, Radev, and Fried 2002). Metacercariae can live up to four years in fish (Elliott and Russert 1949). It transforms into an adult form when the final host consumes amphibians or fish (Bruni and Angelini 2016).

In addition to pathological changes such as deformation and bleeding in the muscles and gills of fish, it causes erythema in the tissues (Yarsan 2020). It affects economically by causing carcass loss due to lesions in fish muscle (Sutili, Gressler, and de Pelegrini 2014). Although it does not pose a big problem except for severe infections, the visible lesions reduce the marketability of the fish. *Clinostomum* spp. is of zoonotic importance and causes "Halzoun syndrome" in people consuming undercooked or raw fish infected with metacercariae (Lee et al. 2017). There is no therapeutic drug used in humans. General anaesthesia is used for mechanical removal. Since the parasite adheres tightly to the mucosa and its rapid movement will prevent the intervention, it is recommended to apply lidocaine on it to reduce the movement of the parasite and to paralyse it (Kitagawa et al. 2003; Sutili, Gressler, and de Pelegrini 2014). The aim of controlling parasitic diseases is to break the cycle. Therefore, waterfowl and water snails must be dealt with. Medicines containing praziquantel can be used in the treatment. Deaths may occur not due to the parasite itself, but to the secreted toxin or secondary infections to places with impaired tissue integrity. At necropsy, cysts can be seen in the skin and muscle with the naked eye. It can be easily recognized due to the size and colour of the cyst. The metacercariae in the cysts are examined under a microscope (Yarsan 2020). Species identification can be made by molecular studies.

Eubothriosis

Eubothrium crassum and *E. salvelini* are the most abundant species in this genus. Salmonid species serve as intermediate or final hosts. *E. salvelini* has been reported in rainbow trout, char, brook trout, Pacific, and Atlantic salmon, and *E. crassum* in brown trout, rainbow trout, Atlantic salmon, vendace, and Danube salmon (Bruno, Noguera, and Poppe 2013). The segmentation is fully formed and can reach 10-60 cm in length and even up to 1 m, depending on the species. The scolex is elongated or spherical in shape and it has bothria and an apical disc. The shape of the disc and the number of slots and grooves on it vary according to the species (Yarsan 2020). In the life cycle, eggs from infected fish are released into the water and these eggs are swallowed by copepods, where procercoid larvae are formed (Saksvik et al. 2001). Plerocercoid larvae occur as a result of ingestion of copepods by the final host fish (Kennedy 1996). In another case, immature plerocercoid results from ingestion by the paratenic host. As the last host eats this paratenic host, the plerocercoid larva develops in the final host (Ruud 2019). It lays its eggs in the wild when copepods are most abundant in order to complete the cycles and reach the intermediate host quickly (Ruud 2019). Copepods are most abundant in coastal areas between May and July (Deschutter et al. 2019). These parasites, which are not very pathogenic, cause growth retardation, feed utilization, and weakening of fish in farms (Yarsan 2020). Since they can extend up to 1m in the intestines, they cause obstruction. Diagnosis is made by finding mature parasites in the intestine at necropsy.

Eustrongylidosis

Eustrongylides spp are nematodes confirmed by samples taken from the stomachs of ciconiiformes and waterfowl brought from Egypt and Sudan by Jägerskiöld in 1909 (Jägerskiöld 1909). The final larval stage in the development of *Eustrongylides* sp. becomes pathogenic for fish (Sloboda et al. 2010). Its larvae in fish have been reported in 17 fish sets worldwide (Spalding and Forrester 1993). Eustrongylidosis is found in every climate zone except arctic, subarctic, and Antarctic. It is more common in temperate continental, tropical and subtropical regions (Mierzejewska et al. 2012). The parasite can infect a wide variety of animals, not just intermediate host fish and final host waterfowl. It infects amphibians, reptiles, mammals, and humans (Ljubojevic et al. 2015). While it can cause infection in salt water, it mostly infects fish in fresh water. *Eustrongylides* spp. is important for public health as it causes infection as a result of consuming undercooked or raw (Food and Administration 2012). In the life cycle, they transform into L2 and L3 forms in oligochaete worms such as tubifex, limnodrilus, the 1st intermediate host that takes the first period larvae developing in eggs excreted with the feces of the last host waterfowl. In these, they turn into L4 stage when they are taken by fish that feed on plankton such as carp and tench, which is the second intermediate host. If fish containing Eustronglides are eaten by other fish, they serve as paratenic hosts. The parasite leaves the intestine and migrates to the muscles and internal organs, causing similar lesions. In fish-feeding waterfowl, it transforms into the L5 mature form (Yarsan 2020; Honcharov et al. 2022). The larvae of this parasite are parasites of easily distinguishable colour and size in fish tissues (Regulation 2002). The image of *Eustrongylides excisus* is shown in Figure 11.

Parasite larvae are encapsulated and inflammation around the larvae is evident, especially in the mesentery and muscles (Goncharov et al. 2018).

Histopathology shows necrosis, degeneration, macrophage aggregation, and multifocal fibroepithelial granulomas in the areas of infection (Dezfuli et al. 2015). Melanized granulomas concentrated in the muscle tissues of the fish are noticed by fish-eating birds and cause them to be eaten more quickly (Coyner et al. 2001). Necrosis of the gonads, lysis of eggs, and reductions in the gonadosomatic index are seen in infected and non-infected (Kaur, Shrivastav, and Qureshi 2013). *Eustrongylides* spp. is a zoonotic parasite, and perforation and gastritis occur in the gastrointestinal tract as a result of consuming meat infected with these fish that has not been sufficiently heat-treated (Deardorff and Overstreet 1991).

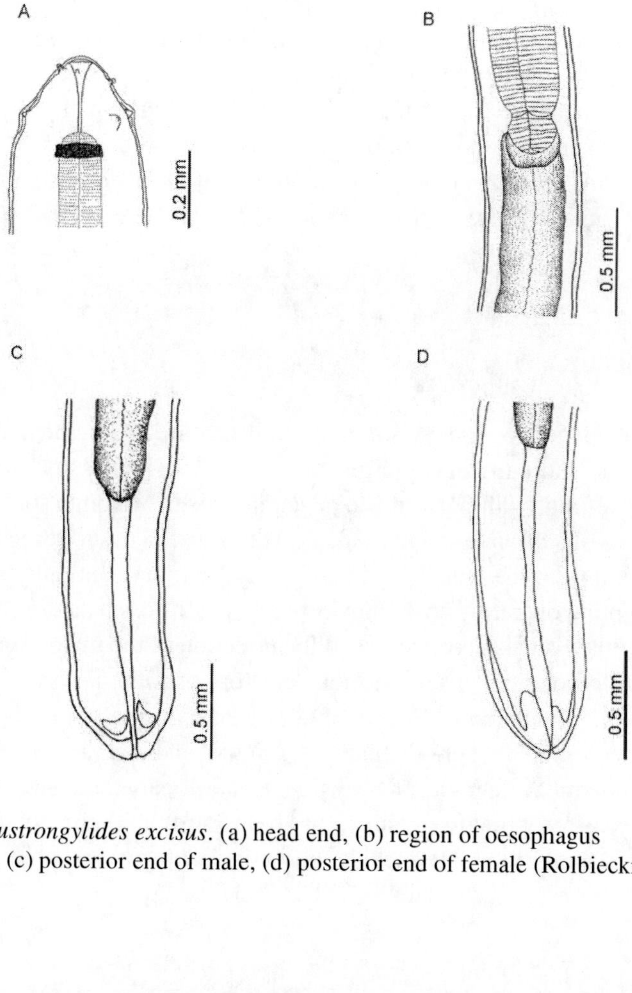

Figure 11. *Eustrongylides excisus*. (a) head end, (b) region of oesophagus and intestine, (c) posterior end of male, (d) posterior end of female (Rolbiecki, 2010).

Myxobolus

Myxobolus Bütschli is the most species-rich genus in the 1882 myxozoan (Zhao et al. 2013). *Myxobolus* contains about 905 nominal species in the genus Bütschli (Eiras, Saraiva, and Cruz 2014). It has a very wide distribution area, with its spread in more than 60 countries, including North America, South America, Asia, Africa, Europe, and Oceania (Eiras, Saraiva, and Cruz 2014). Some species are highly pathogenic for wildlife and farm fish, for example *M. cerebralis* Hofer is one of the important species among 1903 Myxozoa species and is the causative agent of rotational disease in fish (Feist 2008). While the majority of fish species infected by *Myxobolus* parasites live in fresh and brackish waters, a small part of them live in seas (Lom and Dyková 2013). Spores of the genus *Myxobolus* are ellipsoidal, oval, or round in appearance and their shell valves are smooth. There are often equally pear-shaped or sometimes unequal polar capsules. Parasites in the genus *Myxobolus* are localized in different tissues and organs such as liver, kidney, gill, gonad, intestine, and skin as cysts (plasmodia) of different sizes (Feist and Longshaw 2006).

Conclusion

Common parasitic diseases in rainbow trout are mentioned. Parasites can cause direct mortality in fish farms. In cases where it does not cause mortality, a decrease in growth performance occurs in fish. Although it does not cause death in fish, the lesions it creates in the muscle tissue directly affect the marketability of the fish. This, in turn, may limit the continuity of businesses by disrupting operating costs and economic profit-loss balance. Damage to the tissue of the fish also contributes to the infection of the microorganisms in the aquatic environment. The best treatment for parasitic diseases is prevention. Fish infected with parasites should not be kept in pools and cages where the fish are not infected. Periodic parasite controls and disinfection are important. In the treatment, the use of drugs against ectoparasites, endoparasites and protozoa gives effective results. In the control of helminth diseases, the important thing is the intermediate host control.

References

Abdullah, Shamall. 2013. *Study on the Parasites of some Fishes from Darbandikhan Lake in Kurdistan Region*, Iraq.

Adams, MB, and BF Nowak. 2004. Experimental amoebic gill disease of Atlantic salmon, Salmo salar L.: further evidence for the primary pathogenic role of Neoparamoeba sp. (Page, 1987). *Journal of Fish Diseases* 27 (2):105-113.

Akıncı, Ahmet Gündüz. 1999. *Uluabat (Apolyont) Gölü tahta balıkları (blicca björkna L. 1758, cyprinidae)'nda plathelminth parazitlerin tesbitine yönelik araştırmalar [Investigations on detection of flathelminth parasites in plankton (blicca björkna L. 1758, cyprinidae) Lake Uluabat (Apolyont)]*, Uludağ Üniversitesi.

Awakura, T, H Kojima, and H Tanaka. 1984. Studies on parasites of masu salmon, Oncorhynchus masou, 7: Costiasis of pond-reared masu salmon fry. *Scientific Reports of the Hokkaido Fish Hatchery (Japan).*

Bakke, Tor A, Phil D Harris, and Jo Cable. 2002. Host specificity dynamics: observations on gyrodactylid monogeneans. *International journal for parasitology* 32 (3):281-308.

Balta, F, and Z Dengiz Balta. 2018. Gökkuşağı alabalığı (*Oncorhynchus mykiss*) yavrularında görülen Hexamitozis' in tanı ve tedavisi [Diagnosis and treatment of hexamitosis in rainbow trout (*Oncorhynchus mykiss*) fry]. *Anadolu Çev ve Hay Dergisi 2018a* 1:4-8.

Becker, C Dale. 1977. Flagellate parasites of fish. *Parasitic protozoa* 1:357-416.

Bostock, John, Brendan McAndrew, Randolph Richards, Kim Jauncey, Trevor Telfer, Kai Lorenzen, David Little, Lindsay Ross, Neil Handisyde, and Iain Gatward. 2010. Aquaculture: global status and trends. *Philosophical Transactions of the Royal Society B: Biological Sciences* 365 (1554):2897-2912.

Bridle, AR, PBB Crosbie, K Cadoret, and BF Nowak. 2010. Rapid detection and quantification of *Neoparamoeba perurans* in the marine environment. *Aquaculture* 309 (1-4):56-61.

Bruni, Giacomo, and Claudio Angelini. 2016. No short term effect of *Clinostomum complanatum* (Trematoda: Digenea: Clinostomatidae) on survival of *Triturus carnifex* (Amphibia: Urodela: Salamandridae). *Acta Herpetologica* 11 (1):75-79.

Bruno, David W, Patricia A Noguera, and Trygve T Poppe. 2013. *A colour atlas of salmonid diseases*. Vol. 91: Springer Science & Business Media.

Buchmann, K, J Bresciani, and C Jappe. 2004. Effects of formalin treatment on epithelial structure and mucous cell densities in rainbow trout, *Oncorhynchus mykiss* (Walbaum), skin. *Journal of Fish Diseases* 27 (2):99-104.

Buchmann, Kurt. 2020. Immune response to *Ichthyophthirius multifiliis* and role of IgT. *Parasite immunology* 42 (8):e12675.

Bullard, Stephen A, and Robin M Overstreet. 2008. Digeneans as enemies of fishes. In *Fish Diseases (2 Vols.)*: CRC Press.

Bullock, AM, and DA Robertson. 1982. A note on the occurence of *Ichtyobodo necator* (Henneguy, 1883) in a wild population of juvenile plaice, *Pleuronectes platessa* L. *Journal of Fish Diseases* 5 (6):531-533.

Busch, S, Inger Dalsgaard, and K Buchmann. 2003. Concomitant exposure of rainbow trout fry to *Gyrodactylus derjavini* and *Flavobacterium psychrophilum*: effects on infection and mortality of host. *Veterinary parasitology* 117 (1-2):117-122.

Buschkiel, Alfred L. 1910. Beiträge zur kenntnis des *Ichthyophthirius multifiliis* Fouquet [Contributions to the knowledge of *Ichthyophthirius multifiliis* Fouquet]. *Archiv Für Protistenkunde* 21:61-102.

Calhoun, DM, KL Leslie, TB Riepe, TJ Achatz, T McDevitt-Galles, VV Tkach, and PTJ Johnson. 2020. Patterns of *Clinostomum marginatum* infection in fishes and amphibians: integration of field, genetic, and experimental approaches. *Journal of helminthology* 94.

Carneiro, PCF, ML Martins, and EC Urbinati. 2002. Effect of sodium chloride on physiological responses and the gill parasite, *Piscinoodinium* sp., in matrinxã, *Brycon cephalus*, (Teleostei: Characidae) subjected to transport stress. *Journal of Aquaculture in the Tropics* 17 (4):337-348.

Coyne, Robert S, Linda Hannick, Dhanasekaran Shanmugam, Jessica B Hostetler, Daniel Brami, Vinita S Joardar, Justin Johnson, Diana Radune, Irtisha Singh, and Jonathan H Badger. 2011. Comparative genomics of the pathogenic ciliate *Ichthyophthirius multifiliis*, its free-living relatives and a host species provide insights into adoption of a parasitic lifestyle and prospects for disease control. *Genome biology* 12 (10):1-26.

Coyner, Donald F, Sarah R Schaack, Marilyn G Spalding, and Donald J Forrester. 2001. Altered predation susceptibility of mosquitofish infected with *Eustrongylides ignotus*. *Journal of Wildlife Diseases* 37 (3):556-560.

Cross, DG. 1972. A review of methods to control ichthyophthiriasis. *The Progressive Fish-Culturist* 34 (3):165-170.

Crowden, AE, and DM Broom. 1980. Effects of the eyefluke, *Diplostomum spathaceum*, on the behaviour of dace (*Leuciscus leuciscus*). *Animal Behaviour* 28 (1):287-294.

D'Agaro, Edo, PierPaolo Gibertoni, and Stefano Esposito. 2022. Recent Trends and Economic Aspects in the Rainbow Trout (*Oncorhynchus mykiss*) Sector. *Applied Sciences* 12 (17):8773.

Deardorff, Thomas L, and Robin M Overstreet. 1991. Seafood-transmitted zoonoses in the United States: the fishes, the dishes, and the worms. In *Microbiology of marine food products*: Springer.

Deschutter, Yana, Karel De Schamphelaere, Gert Everaert, Christoph Mensens, and Marleen De Troch. 2019. Seasonal and spatial fatty acid profiling of the calanoid copepods *Temora longicornis* and *Acartia clausi* linked to environmental stressors in the North Sea. *Marine environmental research* 144:92-101.

Dezfuli, Bahram S, Maurizio Manera, Massimo Lorenzoni, Flavio Pironi, Andrew P Shinn, and Luisa Giari. 2015. Histopathology and the inflammatory response of European perch, *Perca fluviatilis* muscle infected with *Eustrongylides* sp. (Nematoda). *Parasites & Vectors* 8 (1):1-9.

Eiras, Jorge Costa, Aurélia Saraiva, and Cristina Cruz. 2014. Synopsis of the species of Kudoa Meglitsch, 1947 (Myxozoa: Myxosporea: Multivalvulida). *Systematic Parasitology* 87 (2):153-180.

Elliott, AM, and LR Russert. 1949. Some condition characteristics of a yellow perch population heavily parasitized by *Clinostomum marginatum. The Journal of Parasitology* 35 (2):183-190.

Erer, Hüdaverdi. 2002. *Balık Hastalıkları* [*Fish Diseases*]: Selçuk Üniversitesi. ISBN: 975-97351-1-3

Evans, Wallace A. 1974. The histopathology of cutthroat trout experimentally infected with the blood fluke *Sanguinicola klamathensis. Journal of Wildlife Diseases* 10 (3):243-248.

FAO. *The State of World Fisheries and Aquaculture. Towards Blue Transformation. 2022* 2022. Available from www.fao.org.

Feist, Stephen W. 2008. Myxozoan diseases. In *Fish Diseases (2 Vols.)*: CRC Press.

Feist, Stephen W, and Matt Longshaw. 2006. Phylum myxozoa. *Fish diseases and disorders* 1:230-296.

Food, and Drug Administration. 2012. *Eustrongylides* species. *Bad bug book, foodborne pathogenic microorganisms and natural toxins*:158-162.

Francis-Floyd, Ruth, and Peggy Reed. 1991. *Ichthyophthirius multifiliis (white spot) infections in fish*: Florida Cooperative Extension Service, Institute of Food and Agricultural.

Fromm, Bastian. 2014. *Occurrence and phylogenetic implications of microRNAs in the fish parasite Gyrodactylus salaris* (Platyhelminthes: Neodermata: Monogenea) and related species.

Goncharov, SL, NM Soroka, IY Pashkevich, AI Dubovyi, and Alla Bondar. 2018. *Infection of predatory fish with larvae of Eustrongylides excisus (Nematoda, Dioctophymatidae) in the Delta of the Dnipro River and the Dnipro-Buh Estuary in Southern Ukraine*.

Guidelli, G., Isaac, A., & Pavanelli, G. (2002). *Sanguinicola platyrhynchi* n. sp.(Digenea: Sanguinicolidae) parasite of visceral cavity of *Hemisorubim platyrhynchos* (Valenciennes, 1840)(Pisces: Pimelodidae) from the floodplain of the upper Paraná River, Brazil. *Brazilian Journal of Biology, 62*, 801-806.

Gustinelli, A., Caffara, M., Florio, D., Otachi, E., Wathuta, E., & Fioravanti, M. (2010). First description of the adult stage of *Clinostomum cutaneum* Paperna, 1964 (Digenea: Clinostomidae) from grey herons *Ardea cinerea* L. and a redescription of the metacercaria from the Nile tilapia *Oreochromis niloticus niloticus* (L.) in Kenya. *Systematic parasitology*, 76, 39-51. doi:10.1007/s11230-010-9231-5

Hausmann, Katharine, and Molly Gleeson. 2022. Hadfield C, Clayton L: Clinical Guide to Fish Medicine. Wiley Blackwell. Published August 2021: 624 pages. *Journal of Exotic Pet Medicine* 40:83.

Heinecke, Rasmus D, and Kurt Buchmann. 2009. Control of *Ichthyophthirius multifiliis* using a combination of water filtration and sodium percarbonate: dose-response studies. *Aquaculture* 288 (1-2):32-35.

Honcharov, SL, NM Soroka, MV Galat, OV Zhurenko, AI Dubovyi, and VI Dzhmil. 2022. (Nematoda: Dioctophymatidae): epizootology and special characteristics of the development biology. *Helminthologia* 59 (2):127-142.

Isaksen, Trond. 2013. *Ichthyobodo infections on farmed and wild fish - Methods for detection and identification of Ichthyobodo spp*.

Jägerskiöld, Leonard Axel. 1909. *Zur Kenntnis der Nematoden-Gattungen Eustrongilides und Hystrichis* [*About the nematode genera Eustrongilides and Hystrichis*]. Vol. 2: Akademische Buchdruckerei.

Jeney, Galina. 2017. *Fish diseases: prevention and control strategies*: Academic press.

Joyon, Louıs, and Jırı Lom. 1969. Etude cytologique, systematique et pathologique *d'Ichtyobodo necator* (Henneguy, 1883) Pinto, 1928 (Zooflagelle) [Cytological, systematic and pathological study of *Ichtyobodo necator* (Henneguy, 1883) Pinto, 1928 (Zooflagella)]. *The Journal of Protozoology* 16 (4):703-719.

Kabata, Zbigniew. 1985. *Parasites and diseases of fish cultured in the tropics*: Taylor & Francis Ltd.

Kanev, I, V Radev, and B Fried. 2002. Family Clinostomidae Lühe, 1901. In *Keys to the Trematoda: Volume 1*: CABI Publishing Wallingford UK.

Karvonen, A, O Seppälä, and ET Valtonen. 2004. Eye fluke-induced cataract formation in fish: quantitative analysis using an ophthalmological microscope. *Parasitology* 129 (4):473-478.

Karvonen, Anssi, and David J Marcogliese. 2020. Diplostomiasis (Diplostomum spathaceum and related species). In *Climate change and infectious fish diseases*: CABI Wallingford UK.

Kaur, P., R. Shrivastav, and T. A. Qureshi. 2013. Pathological effects of *Eustrongylides* sp. larvae (Dioctophymatidae) infection in freshwater fish, *Glossogobius giuris* (Ham.) with special reference to ovaries. *J Parasit Dis* 37 (2):245-50.

Kennedy, CR. 1996. Establishment, survival and site selection of the cestode *Eubothrium crassum* in brown trout, Salmo trutta. *Parasitology* 112 (3):347-355.

Kent, ML, J Ellis, JW Fournie, SC Dawe, JW Bagshaw, and DJ Whitaker. 1992. *Systemic hexamitid* (Protozoa: Diplomonadida) infection in seawater pen-reated chinook salmon *Oncorhynchus tshawytscha*. *Diseases of Aquatic Organisms* 14 (2):81-89.

Kiemer, Marius CB, and Kenneth D Black. 1997. The effects of hydrogen peroxide on the gill tissues of Atlantic salmon, *Salmo solar* L. *Aquaculture* 153 (3-4):181-189.

Kitagawa, Norihito, Mayuko Oda, Tadahide Totoki, Seiji Washizaki, Masamichi Oda, and Teiji Kifune. 2003. Lidocaine spray used to capture a live *Clinostomum* parasite causing human laryngitis. *American Journal of Otolaryngology* 24 (5):341-343.

Kum, Cavit, and Selim Sekkin. 2011. The immune system drugs in fish: immune function, immunoassay, drugs. *Recent advances in fish farms*:169-210.

Lahnsteiner, Franz, and Thomas Weismann. 2007. Treatment of ichthyophthiriasis in rainbow trout and common carp with common and alternative therapeutics. *Journal of Aquatic Animal Health* 19 (3):186-194.

Lee, Geum Soo, Sang Wook Park, Jin Kim, Kang Seok Seo, Kyoung Wan You, Jae Hun Chung, Hyeong Chul Moon, and Gun Young Hong. 2017. A case of endoscopically treated laryngopharyngitis resulting from *Clinostomum complanatum* infection. *The Korean Journal of Gastroenterology* 69 (3):177-180.

Lee, JJ, EB Small, DH Lynn, and EC Bovee. 1985. Some techniques for collecting, cultivating and observing protozoa. *An illustrated guide to the Protozoa*:1-7.

Lieke, T., Meinelt, T., Hoseinifar, S. H., Pan, B., Straus, D., & Steinberg, C. (2019). Sustainable aquaculture requires environmental-friendly treatment strategies for fish diseases. *Reviews in Aquaculture*, 12. doi:10.1111/raq.12365

Ljubojevic, Dragana, Nikolina Novakov, Vesna Djordjevic, Vladimir Radosavljevic, Milos Pelic, and Miroslav Cirkovic. 2015. Potential parasitic hazards for humans in fish meat. *Procedia Food Science* 5:172-175.

Lom, J, and G Schubert. 1983. Ultrastructural study of *Pisdnoodinium pillulare* (Schäperclaus, 1954) Lom, 1981 with special emphasis on its attachment to the fish host. *Journal of Fish Diseases* 6 (5):411-428.

Lom, Jiří, and Iva Dyková. 2013. Myxozoan genera: definition and notes on taxonomy, life-cycle terminology and pathogenic species. *Folia parasitologica* 53 (1):1-36.

Marcos-López, Mar, and Hamish D Rodger. 2020. Amoebic gill disease and host response in Atlantic salmon (*Salmo salar* L.): A review. *Parasite immunology* 42 (8):e12766.

Matthews, RA. 2005. *Ichthyophthirius multifiliis* Fouquet and ichthyophthiriosis in freshwater teleosts. *Advances in parasitology* 59:159-241.

McAllister, CHRIS T. 1990. Metacercaria of *Clinostomum complanatum* (Rudolphi, 1814) (Trematoda: Digenea) in a Texas salamander, *Eurycea neotenes* (Amphibia: Caudata), with comments on *C. marginatum* (Rudolphi, 1819). *Journal of the Helminthological Society of Washington* 57 (1):69-71.

Meyer, Fred P, and GL Hoffman. 1974. *Parasites of freshwater fishes; II, Protozoa 3. Ichthyophthirius multifiliis*.

Mierzejewska, Katarzyna, Yuriy Kvach, Małgorzata Woźniak, Anna Kosowska, and Janina Dziekońska-Rynko. 2012. Parasites of an Asian fish, the Chinese sleeper *Perccottus glenii*, in the Włocławek Reservoir on the Lower Vistula River, Poland: in search of the key species in the host expansion process. *Comparative parasitology* 79 (1):23-29.

Morrison, RN, PBB Crosbie, MT Cook, MB Adams, and BF Nowak. 2005. Cultured gill-derived *Neoparamoeba pemaquidensis* fails to elicit amoebic gill disease (AGD) in Atlantic salmon Salmo salar. *Diseases of aquatic organisms* 66 (2):135-144.

Munday, BL. 1986. *Diseases of salmonids. Paper read at Proceedings of the workshop on diseases of Australian fish and shellfish*.

Munday, BL, D Zilberg, and V Findlay. 2001. Gill disease of marine fish caused by infection with *Neoparamoeba pemaquidensis*. *Journal of Fish Diseases* 24 (9):497-507.

Noga, Edward J. 2010. *Fish disease: diagnosis and treatment*: John Wiley & Sons.

Oldham, Tina, Hamish Rodger, and Barbara F Nowak. 2016. Incidence and distribution of amoebic gill disease (AGD)—An epidemiological review. *Aquaculture* 457:35-42.

Orobets, Vladimir, Ekaterina Lisovets, Sergey Zabashta, and Aleksey Ermakov. 2019. Control of fish parasites in aquaculture. *IOP Conference Series: Earth and Environmental Science* 403 (1):012065.

Paladini, Giuseppe, Andrew P Shinn, Nicholas GH Taylor, James E Bron, and Haakon Hansen. 2021. Geographical distribution of *Gyrodactylus salaris* Malmberg, 1957 (Monogenea, Gyrodactylidae). *Parasites & vectors* 14 (1):1-20.

Pekmezci, Gökmen Zafer, Banu Yardımcı, and Ertan Emek Onuk. 2017. Kızılırmak-Yeşilırmak deltaları arasında kalan tatlı su kaynaklarında yakalanıp insan tüketimine sunulan farklı türdeki balıklarda ülkemiz açısından önemli olan balık kökenli zoonoz parazitlerin araştırılması: Enfeksiyon oranları, Yoğunluk ve Bolluklarının ortaya konulması [Investigation of fish-derived zoonotic parasites, which are important for our country, in different species of fish caught in fresh water resources between the Kızılırmak-Yeşilırmak deltas and offered for human consumption: Infection rates, Density and Abundance]. *Sağlık Bilimleri Dergisi* 26 (1):60-63.

Picon-Camacho, Sara M, Mar Marcos-Lopez, JE Bron, and AP Shinn. 2012. An assessment of the use of drug and non-drug interventions in the treatment of *Ichthyophthirius multifiliis* Fouquet, 1876, a protozoan parasite of freshwater fish. *Parasitology* 139 (2):149-190.

Powell, MD, BF Nowak, and MB Adams. 2002. Cardiac morphology in relation to amoebic gill disease history in Atlantic salmon, *Salmo salar* L. *Journal of Fish Diseases* 25 (4):209-215.

Poynton, Sarah L, and Carol M Morrison. 1990. Morphology of diplomonad flagellates: *Spironucleus torosa* n. sp. from Atlantic cod *Gadus morhua* L., and haddock *Melanogrammus aeglefinus* (L.) and *Hexamita salmonis* Moore from brook trout *Salvelinus fontinalis* (Mitchill). *The Journal of protozoology* 37 (5):369-383.

Regulation, EU. 2002. No 178/2002 laying down the general principles and requirements of food law, establishing the European Food Safety Authority and laying down procedures in matters of food safety. *Official Journal L* 31:01-02.

Roberts, Ronald J. 2012. *Fish pathology*: John Wiley & Sons.

Rodger, HD. 2014. Amoebic gill disease (AGD) in farmed salmon (*Salmo salar*) in Europe. *Fish Veterinary Journal* (14):16-27.

Rodger, HD, and JF McArdle. 1996. An outbreak of amoebic gill disease in Ireland. *Veterinary record* 139:348-348.

Rolbiecki, L. (2010). Diversity of metazoan parasite communities in selected fish species from water basins with different degrees of anthropogenic stress. *Oceanological and Hydrobiological Studies*, 39, 3-150.

Rozas-Serri, Marco. 2019. Gill diseases in marine salmon aquaculture with an emphasis on amoebic gill disease. *CABI Reviews* (2019):1-15.

Ruud, Kristian. 2019. *Infection dynamics of marine Eubothrium sp. (Cestoda) in farmed Atlantic salmon*, The University of Bergen.

Saksvik, Monica, Are Nylund, Frank Nilsen, and Kjartan Hodneland. 2001. Experimental infection of Atlantic salmon (*Salmo salar*) with marine *Eubothrium* sp. (Cestoda: Pseudophyllidea): observations on the life cycle, aspects of development and growth of the parasite. *Folia parasitologica* 48 (2):118-126.

Seppälä, Otto, Anssi Karvonen, and E Tellervo Valtonen. 2008. Shoaling behaviour of fish under parasitism and predation risk. *Animal Behaviour* 75 (1):145-150.

Shaharom-Harrison, FM, IG Anderson, AZ Siti, Noor AM Shazili, KJ Ang, and TI Azmi. 1990. Epizootics of Malaysian cultured freshwater pond fishes by *Piscinoodinium pillulare* (Schaperclaus 1954) Lom 1981. *Aquaculture* 86 (2-3):127-138.

Shwani, Abdulkarim, and Shamall Abdullah. 2010. Endoparasites of the Asian Catfish Silurus Triostegus (Heckel, 1843) from Greater Zab River-Kurdistan Region-Iraq. *J. Duhok Univ. (Special Issue)* 13:172-179.

Simsek, Emrah, Alparslan Yildirim, Erdal Yilmaz, Abdullah Inci, Onder Duzlu, Zuhal Onder, Arif Ciloglu, Gamze Yetismis, and Gokmen Zafer Pekmezci. 2018. Occurrence and molecular characterization of *Clinostomum complanatum* (Trematoda: Clinostomidae) in freshwater fishes caught from Turkey. *Parasitology research* 117 (7):2117-2124.

Skrudland, A. 1987. An outbreak of *Ichthyobodo necator* in salmon fry. *Norsk Veterinaertidsskrift* 99 (10):729-730.

Sloboda, Michal, Andrei Daniel Mihalca, Istvan Falka, Klára J Petrželková, Martin Carlsson, Ioan Ghira, and David Modrý. 2010. Are gobiid fish more susceptible to predation if parasitized by *Eustrongylides excisus*? An answer from robbed snakes. *Ecological research* 25 (2):469-473.

Smyth, James Desmond, and MM Smyth. 1980. *Frogs as host-parasite systems 1*: The Macmillan Press Ltd.

Spalding, Marilyn G, and Donald J Forrester. 1993. Pathogenesis of *Eustrongylides ignotus* (Nematoda: Dioctophymatoidea) in ciconiiformes. *Journal of Wildlife Diseases* 29 (2):250-260.

Steinum, Terje, A Kvellestad, LB Rønneberg, H Nilsen, A Asheim, K Fjell, SMR Nygård, AB Olsen, and OB Dale. 2008. First cases of amoebic gill disease (AGD) in Norwegian seawater farmed Atlantic salmon, *Salmo salar* L., and phylogeny of the causative amoeba using 18S cDNA sequences. *Journal of fish diseases* 31 (3):205-214.

Subramani, Parasuraman A, and R Dinakaran Michael. 2017. Prophylactic and prevention methods against diseases in aquaculture. In *Fish diseases*: Elsevier.

Sutili, Fernando Jonas, Luciane Tourem Gressler, and Luis Fernando Vilani de Pelegrini. 2014. *Clinostomum complanatum* (Trematoda, Digenea): a parasite of birds and fishes with zoonotic potential in southern Brazil. A Review. *Revista Brasileira de Higiene e Sanidade Animal: RBHSA* 8 (1):99-114.

Taylor, Richard S, Warren J Muller, Mathew T Cook, Peter D Kube, and Nicholas G Elliott. 2009. Gill observations in Atlantic salmon (*Salmo salar*, L.) during repeated amoebic gill disease (AGD) field exposure and survival challenge. *Aquaculture* 290 (1-2):1-8.

Valdenegro-Vega, Victoria A, Mathew Cook, Philip Crosbie, Andrew R Bridle, and Barbara F Nowak. 2015. Vaccination with recombinant protein (r22C03), a putative attachment factor of *Neoparamoeba perurans*, against AGD in Atlantic salmon (*Salmo salar*) and implications of a co-infection with *Yersinia ruckeri*. *Fish & shellfish immunology* 44 (2):592-602.

Vannetti, S. M., Wynne, J. W., English, C., Huynh, C., Knüsel, R., de Sales-Ribeiro, C., . . . Schmidt-Posthaus, H. (2023). Amoeba species colonizing the gills of rainbow trout (*Oncorhynchus mykiss*) in Swiss aquaculture. *Journal of fish diseases*.

Verner-Jeffreys, David, and Nick Taylor. 2015. *Review of Freshwater Treatments Used in the Scottish Freshwater Rainbow Trout Aquaculture Industry*.

Vickerman, K. 1990. Phylum Zoomastigina Class Diplomonadida. *Handbook of Protoctista. Jones & Bartlett, Boston*:200-210.

Woo, Patrick TK, and Kurt Buchmann. 2012. *Fish parasites: pathobiology and protection*: Cabi.

Woo, Patrick TK, and John F Leatherland. 2006. *Fish diseases and disorders*: CABI.

Yarsan, Ender. 2020. *Balık Hekimligi*. ISBN: 978-975-277-809-2.

Young, ND, I Dyková, BF Nowak, and RN Morrison. 2008. Development of a diagnostic PCR to detect *Neoparamoeba perurans*, agent of amoebic gill disease. *Journal of Fish Diseases* 31 (4):285-295.

Young, Neil D, Iva Dyková, Kevin Snekvik, Barbara F Nowak, and Richard N Morrison. 2008. Neoparamoeba perurans is a cosmopolitan aetiological agent of amoebic gill disease. *Diseases of aquatic organisms* 78 (3):217-223.

Zhao, YJ, NN Li, FH Tang, and JL3825597 Dong. 2013. Remarks on the validity of *Myxobolus ampullicapsulatus* and *Myxobolus honghuensis* (Myxozoa: Myxosporea) based on SSU rDNA sequences. *Parasitology Research* 112 (11):3817-3823.

Biogrpahical Sketch

Mustafa ÖZ

Affiliation: Aksaray University

Education: PhD

Business Address: Department of Fisheries and Diseases, Faculty of Veterinary Medicine, Aksaray University, Türkiye

Research and Professional Experience: Fish feeding, Aquaculture, Fish diseases, Fish processing technology.

Professional Appointments:

- Director of Eskil Vocational High School
- Vice Dean of the Faculty of Veterinary Medicine
- Head of preclinical sciences department
- Head of Department of Fisheries and Diseases

Honors:

- Academic Performance Award, Aksaray University, Türkiye.
- Scientific Publication Award, the Scientific and Technological Research Council of Türkiye.

Publications from the Last 3 Years:

Book Chapters

Öz, M., Ucak, İ., & Nayik, G. A. (2022). PUFA and MUFA. In *Nutraceuticals and Health Care* (pp. 199-215). Academic Press.

Uçak, İ., Öz, M., & Maqsood, S. (2019). Products based on omega-3 polyunsaturated fatty acids and health effects. In *The Role of Alternative and Innovative Food Ingredients and Products in Consumer Wellness* (pp. 197-212). Academic Press.

Öz, M. 2020. Black cumin (*Nigella sativa* l.) seed and oil: composition, uses and health effects. *Theory and Research in Agriculture, Forestry and Aquaculture Sciences II*, 131.

ÖZ, M., & Üstüner, E. *Ot sazani (ctenopharyngodon idella, valenciennes 1844) yetiştiriciliği. Ilıman İklim Balıkları Yetiştiriciliği* [*Cultivation of grass carp (ctenopharyngodon idella, valenciennes 1844). Temperate Climate Fish Breeding*]. Editör, Suat Dikel. Akademisyen Kitapevi. 155-167.

Öz, M., & Üstüner, E. (2021). Sazan (Cyprinus carpio Linnaeus, 1758) Yetiştiriciliği. *Ilıman İklim Balıkları Yetiştiriciliği* [Carp (Cyprinus carpio Linnaeus, 1758) *Breeding. Temperate Climate Fish Breeding*]. Editör, Suat Dikel. Akademisyen kitabevi. 113-130.

Öz, M. 2022. *Baliklarda Büyümenin Parametrik Değerlendirilmesi* [*Parametric Evaluation of Growth in Fish*]. Editörs, Suat DİKEL, Ibrahim Demirkale. iksad publishing house. 119-142.

Öz, M. 2022. *Üretim Işletmesi Kurulumu* [*Production Business Setup*]. Editörs, Suat Dikel, Ibrahim Demirkale. iksad publishing house. 55-74.

Articles

Öz, M. & Dikel, S. (2022). Effect of garlic (*Allium sativum*)-supplemented diet on growth performance, body composition and fatty acid profile of rainbow trout (*Oncorhynchus mykiss*). *Cellular and Molecular Biology*, *68* (1), 217-225.

Dikel, S., Özşahinoğlu, I., Öz, M., & Demirkale, İ. (2022). Growth Performances of Shibot Fish (*Tor grypus*) Fry Fed with Diets Containing Different Protein Levels. *Turkish Journal of Agriculture-Food Science and Technology*, *10* (5), 933-940.

Öz, M. & Bağdatli, İ. (2021). Farklı oranlarda limon suyu ile marine edilmiş sazan etinin (*Cyprinus carpio*) duyusal parametrelerinin belirlenmesi [Determination of sensory parameters of carp meat (*Cyprinus carpio*) marinated with different proportions of lemon juice]. *Avrupa Bilim ve Teknoloji Dergisi*, (31), 681-685.

Öz, M. Tatil, T., & Dikel, S. (2021). Effects of boric acid on the growth performance and nutritional content of rainbow trout (*Oncorhynchus mykiss*). *Chemosphere*, *272*, 129895.

Öz, M., İnanan, B., & Dikel, S. (2021). Yem Kaynaklı Borun Gökkuşağı Alabalığı (*Oncorhynchus mykiss*) Yağ Asidi Profiline Etkisi [Effect of Feed Source Boron on Fatty Acid Profile of Rainbow Trout (*Oncorhynchus mykiss*)]. *Avrupa Bilim ve Teknoloji Dergisi*, (31), 188-192.

Öz, M., Yavuz, O., & Bolukbas, F. (2020). Histopathology changes in the rainbow trout (*Onchorhyncus mykiss*) consuming boric acid supplemented fish fodder. *Journal of Trace Elements in Medicine and Biology*, *62*, 126581.

Öz, M., Karaşahin, T., Aksoy, N. H., Inanan, B. E., & Dikel, S. (2020). Harmful effects of dietary supplementation of boron on blood parameters of Rainbow Trout (*Oncorhynchus mykiss*). *Journal of the Hellenic Veterinary Medical Society*, *71* (2), 2227-2234.

Öz, M. (2019). Effects of habitats and feeding patterns on fatty acid profile of rainbow trout (*Oncorhynchus mykiss*). *Eurasian Journal of Food Science and Technology*, *3* (2), 34-39.

Öz, M. (2019). Use of Herbal Food Additives in Fisheries. *Eurasian Journal of Agricultural Research*, *3* (2), 67-74.

Chapter 6

Nutritional Composition, Fatty Acid Profile and Shelf Life of Rainbow Trout (*Oncorhynchus mykiss*)

Mustafa Öz[1,*]
and Suat Dikel[2]
[1]Department of Fisheries and Diseases, Faculty of Veterinary Medicine, Aksaray University, Aksaray Turkey
[2]Department of Aquaculture, Faculty of Fisheries, University of Cukurova, Adana, Turkey

Abstract

Rainbow trout (*Oncorhynchus mykiss*), a member of the Salmonidae family, has a high commercial importance and is traded in many parts of the world. Rainbow trout is a member of the Salmonidae family, which is the most widely farmed after the Atlantic salmon. It is widely cultivated due to its high growth performance, easy retrieval of eggs and sperm, and being more resistant to diseases than other trout species. Rainbow trout is a type of fish that is consumed with pleasure because of its delicious meat, rich nutritional content, and fatty acid profile. Conscious consumers prefer fish meat as both a protein source and a source of omega 3. Knowing the nutrient content of the fish is very important in terms of determining the cooking method of the fish, processing the fish meat, and determining the conditions under which it will be stored. One of the most important elements that make fish valuable for human health is the fatty acids in fish oil. One of the most important factors determining the shelf life of fish is the oil content of

[*] Corresponding Author's Email: ozmustafa@aksaray.edu.tr.

In: Advances in Marine Biology. Volume 6
Editors: Adam Kovács and Patrik Nagy
ISBN: 979-8-88697-996-1

fish meat. In this study, a comprehensive study of the nutritional content, fatty acid profile and shelf life of rainbow trout was revealed.

Keywords: fatty acid, rainbow trout, proximate composition, shelf life

Introduction

Fish is one of the foods that significantly affect human health, starting from the mother's womb. In recent years, with consumers' awareness, people have begun to question the nutritional content of the foods they consume. Consumers want to know how much protein and how much fat they consume with every 100 grams of fish they eat, even if this oil contains essential oils and how much. Knowing the nutrient content of fish is very important for us to decide which processing method they will be processed or cooked according to which cooking method.

The rapid decline in natural stocks due to environmental pollution and overfishing pressure has made fish farming one of the fastest-growing food industries. Rainbow trout is the most widely grown fish in the Salmonidae family after Atlantic salmon (*Salmo salar*) and is loved and consumed in many countries.

World aquaculture, which was approximately 30 million tons/per year in 1990, reached 87.5 million tons/per year in 2020. Parallel to this increase in production, rainbow trout production is also increasing rapidly. Rainbow trout production was approximately 340.000 tons in 2000, reaching approximately 740.000 tons in 2020 (FAO, 2022).

With the increase in rainbow trout production, quality parameters, processing, and storage of trout, an important product in international trade, have attracted the attention of many researchers, and much research has been carried out in this field. In this study, the research on the muscle nutrient content, fatty acids, and shelf life of rainbow trout, consumed by consumers and whose cultivation is increasing daily, was meticulously examined and presented.

Proximate Composition and Nutritional Properties of Rainbow Trout

When examining the nutritional content of fish meat, four main components are usually considered. These components are moisture, crude protein, crude fat, and crude ash. Since the carbohydrate content of fish meat is very low, it is often overlooked, and the fifth component, carbohydrates, is not considered (Mishra, 2022). The crude protein, moisture, lipid, and crude ash ratios of rainbow trout are shown in Table 1.

Table 1. Proximate composition of rainbow trout

Constituent	% by weight
Moisture	65.13-77.70
Crude protein	15.67-23.80
Crude fat	1.12-13.50
Ash	0.91-3.74

Table 2. Proximate analysis results of rainbow trout obtained by different researchers

No	Proximate composition (%)				Reference
	Moisture	Protein	Lipid	Ash	
1	74.5	18.70	5.11	1.26	Fallah et al., 2011
2	78.4	19.20	1.12	1.30	Fallah et al., 2011
3	71.70	19.65	4.46	1.33	Sabetian et al., 2012
4	73.38	19.80	3.44	1.35	Gokoglu et al., 2004
5	75.60	17.90	4.30	2.10	Yeşilayer and Genç, 2013
6	76.99	20.83	1.83	0.91	Asghari et al., 2012
7	70.67	21.23	3.88	1.48	Tokur, 2007
8	76.23	20.56	2.67	1.64	Hosseinzadeh et al., 2013
9	77.22	20.47	1.66	1.64	Hosseinzadeh et al., 2013
10	66.3	17.90	13.5	2.3	Güroy et al., 2010
11	71.21	20.28	2.31	1.53	Unusan, 2006
12	74.23	19.70	4.47	1.20	Taşbozan et al., 2016
13	68.79	21.60	7.78	1.37	Taşbozan et al., 2016
14	69.63	22.41	5.90	1.20	Taşbozan et al., 2016
15	69.34	22.45	6.60	1.31	Taşbozan et al., 2016
16	68,90	23.80	5.8	2.5	Chaiyapechara et al., 2003
17	68,10	23.60	6.9	2.3	Chaiyapechara et al., 2003
18	67,60	23.6	7.5	2.2	Chaiyapechara et al.,2003
19	67.80	23.00	7.9	2.1	Chaiyapechara et al., 2003
20	71.28	22.43	6.98	1.40	Nazari et al., 2017
21	74.57	18.79	4.56	1.94	Öz et al., 2015

Table 2. (Continued)

No	Proximate composition (%)				Reference
	Moisture	Protein	Lipid	Ash	
22	72.16	17.31	9.10	1.11	Zare et al., 2021
23	71.64	17.51	8.12	2.44	Zare et al., 2021
24	75.47	16.91	2.22	1.33	Trung et al., 2022
25	76.57	17.88	2.55	1.49	Trung et al., 2022
26	77.56	18.74	4.25	1.33	Trung et al., 2022
27	77.57	19.37	8.59	1.39	Trung et al., 2022
28	71.65	19.60	4.43	1.36	Çelik et al., 2008
29	65.55	17.92	2.34	1.47	Besharat et al., 2021
30	65.54	18.97	1.66	1.43	Besharat et al., 2021
31	65.13	20.53	1.20	1.63	Besharat et al., 2021
32	65.72	15.67	1.86	1.45	Besharat et al., 2021
33	75.70	17.97	4.56	1.33	Öz and Dikel, 2022
34	73.87	20.27	4.17	1.19	Öz and Dikel, 2022
35	77.7	17.96	2.47	1.60	Öz et al., 2021
36	76.25	18.35	3.39	1.90	Öz et al., 2021
37	74.57	18.79	4.56	1.94	Öz et al., 2018
38	71.34	20.37	5.89	2.27	Öz et al., 2018
39	73.62	19.41	3.73	1.78	Öz, 2016
40	75.23	18.73	3.36	1.52	Öz, 2016
41	75.11	19.92	3.46	1.18	Öz and Dikel 2015a
42	73.01	22.33	2.53	1.86	Öz and Dikel 2015b
43	75.69	19.06	3.51	1.62	Öz and Dikel 2015b
44	71.03	19.35	9.26	1.22	Rønsholdt et al., 2004
45	70.74	19.34	9.31	1.20	Rønsholdt et al., 2004
46	70.89	16.69	6.02	1.73	Hunt et al., 2011
47	71.29	19.42	7.33	1.33	Mancini et al., 2018
48	73.21	20.47	3.37	1.49	Wang and Han et al., 2017
49	70.06	23.15	4.14	1.47	Wang and Han et al., 2017
50	72.00	19.30	7.60	1.30	Dumas et al., 2007
51	70.90	15.70	10.40	2.20	Dumas et al., 2007
52	70.34	21.72	4.57	3.74	Rasmussen et al., 2000
53	69,57	22.87	4.16	3.67	Rasmussen et al., 2000
54	70.00	17.20	10.90	1.97	Köse and Yildiz, 2013
55	69.70	17.00	11.90	1.85	Köse and Yildiz, 2013

Since moisture and proteins are the main components of its body, fish is considered a "high-moisture, protein food." Among the four components, protein and ash contents do not vary much, while significant differences are observed in crude fat and moisture content. Generally, there is an inverse relationship between moisture and crude fat content in fish tissue, which together account for about 80% of tissue weight (Mishra, 2022).

The popular consumption of rainbow trout and its easy availability in many markets has led scientists to investigate the nutritional content of this species. The proximate composition of rainbow trout has been studied in different countries, at different times, and by different scientists, and the results obtained are given in Table 2.

Moisture (Water)

The water in fish tissue is the same as normal water and therefore has little nutritional value. However, it is of great importance in the processing of fish. Fish is a high-moisture foodstuff as its moisture content is very high (66-84%). Such high moisture content makes fish a highly perishable food product. In addition, the high moisture content is favorable for many spoilage processes, such as microbial and enzymatic activity. This is because microbes and enzymes require high moisture content for their activity. Therefore, some processing methods, such as drying, salting, and freeze-drying, are based on removing moisture from fish tissue. In freeze preservation, tissue water is also removed in such a way that it is converted into ice, which is unsuitable for microbial or enzymatic activities. When fish is subjected to any external influence (freezing, heating, drying, pH change, or osmotic pressure), the ratio between bound water and free water changes, altering the fresh fish's original properties and negatively affecting its quality (Mishra, 2022).

Proteins

Fish tissue contains many nitrogen-containing substances. These substances are collectively referred to as 'nitrogenous substances.' Some important ones are proteins, free amino acids (FAA), urea, ammonia, trimethylamine oxide (TMAO), creatine, creatinine, etc. The most important nitrogenous substances are proteins. The remainder of the nitrogenous substances that are not part of the protein are collectively referred to as 'non-protein nitrogenous substances' (NPN substances). These include free amino acids (FAA), urea, ammonia, trimethylamine oxide (TMAO), creatine, creatinine, etc. are included.

Fish proteins are divided into three groups according to their location in muscle cells and their solubility in salt solution (Mishra, 2022).

1. Sarcoplasmic proteins (SPP) or water-soluble proteins (WSP)
2. Myofibrillar proteins (MFP) or salt-soluble proteins (SSP)
3. Stroma proteins (SP) or insoluble proteins (IP)

Proteins are important in human health as they affect fish products' sensory properties and nutritional value. Although fish proteins contain all nine essential amino acids (Phenylalanine, histidine, leucine, lysine, threonine, isoleucine, tryptophan, methionine, and valine), fish proteins are more valuable than plant and meat proteins in terms of biological value (BV) and net protein utilization (NPU).

Fish muscle has a high protein content, which is easily digestible, and has a lower collagen content, leading to higher sensitivity than red meat (Savary-Auzeloux et al., l., 2014; Islam & Joadder, 2005).

Lipids

While the crude protein and crude ash ratios of fish do not change much, fish meat's moisture and lipid content vary significantly. The fat content of fish can vary from 0.1% to 22%. Lipids that remain solid or semi-solid at room temperature are called "fats" and are predominantly found as "animal fat" in animals. However, lipids found in fish can remain liquid at room temperature after they are removed from the fish. Crude oil in fish and fish products can be obtained by the Soxhlet method. Lipids in fish consist of fatty acids, oils, phospholipids, cholesterol, hydrocarbons, fat-soluble vitamins, and pigments (Mishra, 2022).

According to their oil content, they are divided into three groups: lean fish, medium oily fish (semi-oily fish), and oily fish.

1. Lean fish: Fish containing less than 2% fat in their meat is called lean fish.
2. Medium-fat fish (Semi-oily fish): The meat of these fish contains between 2% and 5% oil.
3. Oily fish: Fish with more than 5% fat in their meat is called oily fish (Ackman, 1990).

Trout can be considered as oily or medium-oily fish, although the oil content varies according to the season (Huss, 1995).

Fatty Acid Composition of Rainbow Trout

Fatty acids are important structural components with critical roles in metabolic events such as energy storage and transport and regulation of gene functions (Robertson et al., 2013).

They are also necessary for cell membrane structures (Abedi and Sahari, 2014). Fatty acids can be found in the cell in free form and lipid complex (Rustan and Drevon, 2005). All fatty acids are grouped according to their carbon chain lengths, single or double bonds in their structure, and degree of unsaturation (Serpek and Kalaycıoğlu, 2000). Polyunsaturated fatty acids (PUFA) and monounsaturated fatty acids have beneficial and protective effects on human health and development. A large body of experimental evidence suggests that PUFAs and MUFAs play a strong role in suppressing the development of most cancers and reducing the risk of cardiovascular disease. While the main source of n-3 series PUFAs, especially DHA and EPA, is generally found in fish, MUFAs are mostly found in olive oil and canola oil (Öz et al., 2022; Ucak et al., 2019).

Some of the results obtained by different researchers at different times regarding the fatty acids of rainbow trout are shown in Table 3.

Table 3. Fatty acid profile of rainbow trout

Fatty acid	Percentage of Total Fatty Acids by Weight						
14:0	4.2	2.38	4.9	2.99	2.00	5.14	5.9
16:0	22.4	21.3	18.5	11.89	13.3	16.07	26.1
18:0	4.3	6.79	3.8	2.18	5.1	2.41	3.9
SFA	31.8	30.5	27.2	24.88	20.8	23.62	39.1
16:1n-7	5.5	4.16	6.5		6.3	9.71	7.9
18:1n-9	11.6	22.2	24.5	21.44	23.2	21.63	16.2
18:1n-7	2.4	-	-	2.75	2.7	-	-
20:1n-9	3.6	0.67	2.6	-	8.7	10.08	1.9
22:1	3.5	-	0.8	-	-	8.63	1.1
24:1n-9	0.7	-	-	-	-	-	-
MUFA	27.2	29.3	31.4	27.43	46.9	50.49	29.7
18:2n-6	4.5	10.4	11.2	11.77	2.1	3.81	2.6
20:2n-6	0.4	1.03	0.4	-	-	-	-
20:3n-6	0.2	1.17	-	0.23	-	-	-
20:4n-6	0.7	2.56	0.7	0.81	0.7	0.36	0.7
18:3n-3	0.8	-	1.5	3.19	0.4	0.81	0.7
18:4n-3	1.1	-	-	-	0.3	-	1.2
20:4n-3	0.9	-	-	-	0.6	1.19	0.3
20:5n-3	5.9	3.52	4.7	3.42	4.3	4.53	9.3
22:5n-3	1.5	0.85	1.6	-	1.6	1.86	1.2
22:6n-3	24.5	22.7	14.9	9.58	13.2	13.06	18.8
PUFA	41.00	39.20	34.7	30.15	25.1	25.88	35.8
	J. Fonseca-Madrigal et al., 2005	Haliloglu et al., 2004	Yıldız et al., 2018	Öz et al., 2021	Caballero et al., 2002	Johansson et al., 2000	Gomes et al., 1993

According to the results obtained from the research on the fatty acid of rainbow trout, the major fatty acids of rainbow trout are determined and shown in Table 4.

The fat content of rainbow trout is one of the main criteria for assessing its organoleptic quality. When the trout is oily, its meat becomes juicier and softer and is appreciated by consumers. The amount of fat in fish meat is one of the most important factors that determine the shelf life during storage and affects the fish's flavor.

Table 4. Major fatty acids of rainbow trout

Carbon number	Common name
Saturated fatty acids	
C14:0	Myristic acid
C16:0	Palmitic acid
C18:0	Stearic acid
Monounsaturated fatty acids (MUFA)	
C16:1, ω-7	Palmitoleic acid
C18:1, ω-9	Oleic acid
C20: 1, ω-9	Eicosenoic acid
Polyunsaturated fatty acids (PUFA)	
C18: 2, ω-6	Linoleic acid
C18: 3, ω-3	α-Linolenic acid (ALA)
C20: 4, ω-6	Arachidonic acid
C20: 5, ω-3	Eicosapentaenoic acid (EPA)
C22: 5, ω-3	Docosapentaenoic acid (DPA)
C22: 6, ω-3	Docosahexaenoic acid (DHA)

Shelf Life and Shelf Stability of Rainbow Trout

The freshness of fish is closely linked to sensory changes and texture characteristics. Together with pre-mortem muscle biochemistry, post-mortem biochemical processes are directly linked to final quality attributes. Understanding postmortem mechanisms is crucial for accurately controlling fish quality by defining objective indicators. Postmortem softening is one of the most important quality attributes of fish muscle. Loss of freshness results from a complex combination of biochemical, chemical, and physical processes, followed by muscle degradation due to microbiological contamination (Mishra, 2022).

Freshness is the most important feature when assessing fish quality. Microbiological, biochemical, and sensory changes are associated with the deterioration of fish quality during transport and storage (Ehira and Uchiyama, 1986). Given that certain spoilage-causing microorganisms cannot be detected by organoleptic or chemical testing, it is useful to perform microbiological, chemical, and organoleptic analyzes when assessing fish quality (Ryder et al., 1993).

Research Made to Extend the Shelf Life of Rainbow Trout

Studies to extend the shelf life of fish meat can be examined in two groups. In some of the studies, it was aimed to extend the storage period of the fish by improving the muscle biochemistry before death. In contrast, some studies aimed to extend the shelf life by using some antioxidant-containing substances or by trying different processing methods after death.

In a study, the effects of rosemary oil on the freshness indicators, oxidative stability, fatty acid, and biogenic amine content of minced rainbow trout meat were examined. As a result of the research, it was found that rosemary oil improved the freshness indicators, pH of lipids, oxidative stability, and fatty acid profile of minced rainbow trout meat, thereby significantly prolonging its shelf life. The same study also reported that rosemary oil leads to low biogenic amine content, especially putrescine, cadaverine, tyramine, and histamine, in fish meat (Pieretti et al., 2012).

Akhtar et al., (1998) reported that rosemary extract contains a high amount of carnosic acid and phenolic diterpenic substances such as carnosol. In their study, rosemary extract showed a very effective antioxidant effect during storing frozen and refrigerated rainbow trout (*Oncorhynchus mykiss*).

Mexis et al., (2009) investigated the combined effect of oxygen absorber and thyme essential oil (0.4%) on prolonging the shelf life of rainbow trout fillets stored under refrigerator conditions (4°C). Microbiological (total viable count, *Pseudomonas* spp., lactic acid bacteria number, H2S producing bacteria including Shewanella putrefaciens, Enterobacteriaceae and *Clostridium* spp.), physicochemical (pH, PV, TBA, TVBN, and water loss) formed in the product during storage and sensory (smell, taste) changes were examined. As a result of the study, it was determined that the shelf life of the trout fillets was four days for the control group, 7-8 days for the groups containing thyme oil, 13-14 days for the groups containing O2 absorber, and 17 days for the groups containing O2 absorber and thyme oil.

Ojagh et al., (2010) examined the effect of chitosan coating enriched with cinnamon oil on rainbow trout quality during 16 days of cold storage

(4 ± 1ºC). As a result of the research, it was reported that the treatment groups maintained their shelf life until the end of storage (16th day), but the control group had a shelf life of 12 days.

Pyrgotou et al., (2010) investigated the effect of thyme essential oil (0.2% and 0.4%) on salted, modified atmosphere-packaged rainbow trout fillets stored at 4°C for 21 days. It was reported that the number of lactic acid bacteria, H2S-producing bacteria (including Shewanella putrefaciens), Enterobacteriaceae, and *Pseudomonas* spp. were higher in the control group compared to the treatment group.

Frangos et al., (2010) investigated the effects of salt, thyme essential oil (0.2% or 0.4%), and packaging on rainbow trout fillets stored at 4ºC. The number of lactic acid bacteria, H2S-producing bacteria, *Pseudomonas* spp., and Enterobacteriaceae in salted and unsalted fish stored in the normal atmosphere were higher than those in salted fish, fish with oregano oil added or fish stored in vacuum pack conditions without oregano oil added.

Pezeshk et al., (2011) studied the effects of turmeric extract, shallot extract, and their combination on the sensory, chemical, and microbiological quality (*Oncorhynchus mykiss*) of vacuum-packed rainbow trout during 20 days of cold storage (4 ± 1◦C).

Chytiri et al., (2004) studied the effect of filleting on the microbiological, chemical, and sensory characteristics of cultured rainbow trout stored on ice. Pseudomonas, H2S-producing bacteria (including Shewanella putrefaciens), and *Brochothrix thermosphacta* were the dominant bacteria, with lower amounts of Enterobacteriaceae present in whole or filleted rainbow trout meat during 18 days of ice storage.

Investigation of Sensory Characteristics of Rainbow Trout

For sensory analyzes to be applied correctly, the panelists who will decide on the quality of the fish must be experienced. Especially those engaged in fisheries or those working in aquaculture processing technology can make healthy decisions with the experience they have gained over a long time. The decisions made by different people may be different for fish that have the same degree of freshness under the control of their sense of freshness. The psychological state of the person who controls the freshness, whether his body is tired or not, and similar factors may affect the decision to be made. Sensory analysis of fish is usually done in two stages. Sensory evaluation in raw fish meat is sometimes not enough, so it is necessary to perform sensory analysis in raw and cooked fish meat.

Although many methods are used in sensory analysis, one of the most common is the method developed by Bonilla et al., for *Gadus morhua* (*Gadus morhua*), which is shown in Table 5 (Bonilla et al., 2007). Many researchers who made the sensory evaluation of rainbow trout also revised this method and tried to determine the sensory changes during storage (Ozogul et al., 2017; Oz et al., 2017; Oz, 2018; Durmus, 2020; Aksoy and Sezer, 2019; Yavuzer et al., 2007; Fıçıcılar and Gençcelep, 2021).

Table 5. Quality Index Method scheme developed for fresh cod (*Gadus morhua*) fillets with skin (Bonilla et al., 2007)

Quality parameter		Description	Score
Skin	Brightness	Iridescent pigmentation	0
		Rather dull	1
		Dull	2
	Mucus	Uniform, thin, transparent	0
		Little thicker, opaque	1
		Clotted, thick, yellowish	2
Flesh	Texture	Firm	0
		Rather soft	1
		Very soft	2
	Blood	Bright red, not present	0
		Dull red	1
		Shadowy, brown	2
	Odour	Fresh, neutral	0
		Seaweedy, marine, grass	1
		Sour milk	2
		Acetic, ammonia	3
	Colour	White, greyish	0
		Some yellowish, a little pinkish	1
		Yellow, overall pink	2
	Bright	Transparent, bluish	0
		Opaque	1
		Milky	2
	Gaping	No gaping, one longitudinal gaping at the neck part of the fillet	0
		Slight gaping less than 25% of the fillet	1
		Slight gaping, 25–75% of the fillet	2
		Deep gaping or slight gaping over 75% of the fillet	3

Investigation of Chemical Change of Rainbow Trout Meat

To determine chemical changes during the storage of fish, usually, thiobarbituric-acid reactive substance (TBARS) measurement, pH measurement, color measurement, Analysis of the Fatty Acids, Determination of Biogenic Amines, the total volatile base nitrogen (TVB-N), Free fatty acid (FFA) analysis and Peroxide value (PV) are determined. Some chemical changes during the storage of rainbow trout are given in Table 6.

Table 6. Min. max. values of some chemical changes during storage of rainbow trout

Chemical Parameters	Min.	Max	Reference
TBA (mg MA/kg)	0.50	0.98	Özyurt et al., 2015
	2.55	7.57	Erümit and Yıldız, 2022
	0.07	1.21	Mehdizadeh et al., 2019
	0.16	6.24	Dikici et al., 2021
	0.20	2.11	Bazargani-Gilani, 2018
	0.21	1.68	Hassanzadeh et al., 2018
	0.44	1.33	Öz, 2018
	0.50	2.26	Özoğul et al., 2017
	0.47	0.65	Öz et al., 2017
	1.08	9.20	Yıldız and Yangılar, 2016
TVB-N (mg/100 g)	4.89	15.34	Özyurt et al., 2015
	14.95	24.27	Erümit and Yıldız, 2022
	5.6	34.06	Hamzeh and Rezaei, 2012
	11.72	45.23	Mehdizadeh et al., 2019
	3.16	22.1	Dikici et al., 2021
	12.04	40.02	Bazargani-Gilani, 2018
	12.41	32.50	Yıldız and Yangılar, 2016
	12.60	38.00	Pezeshk et al., 2011
	12.81	22.92	Öz, 2018
	13.72	58.18	Özoğul et al., 2017
Peroxide value-PV (meq/kg)	0.43	1.70	Özyurt et al., 2015
	0.54	16.25	Hamzeh and Rezaei, 2012
	2.02	3.92	Yıldız and Yangılar, 2016
	0.13	0.45	Zhaleh et al., 2019
	0.030	0.69	Hassanzadeh et al., 2018
	0.86	6.70	Pezeshk et al., 2011
	6.56	21.57	Öz, 2018
	4.08	18.23	Özoğul et al., 2017
	2.16	14.15	Öz et al., 2017
	0.05	2.11	Mozaffarzogh, et al., 2020

Chemical Parameters	Min.	Max	Reference
pH	6.70	7.04	Peiretti et al., 2012
	6.15	6.49	Erümit and Yıldız, 2022
	5.81	7.03	Özyurt et al., 2015
	6.30	6.53	Dikici et al., 2021
	6.45	6.83	Zhaleh et al., 2019
	3.25	8.24	Hassanzadeh et al., 2018
	6.40	6.72	Öz, 2018
	6.22	6.69	Yıldız and Yangılar, 2016
	6.89	7.21	Özoğul et al., 2017
	6.59	7.23	Öz et al., 2017
Free fatty acid (FFA)- % oleic acid	0.18	0.47	Özyurt et al., 2015
	3.46	11.47	Öz, 2018
	4.92	12.54	Özoğul et al., 2017
	2.30	10.69	Öz et al., 2017
Chemical Parameters	Min.	Max	Reference
	1.50	2.89	Rezaei and Hosseini, 2008
	0.65	3.63	Barkhori-Mehni et al., 2022
	0.56	3.29	Shadman et al., 2017
	1.69	7.08	Ehsani and Jasour, 2012
	2.87	6.78	Nowzari et al., 2013
	0.53	3.61	Hashemi et al., 2022

Microbiological Changes

The subsurface meat of healthy fish is considered sterile and free of bacteria or other microorganisms. However, the saprophytic flora begins to invade the surface flesh after death. The number and diversity of microbes associated with fish vary depending on geographic location, season, and harvest method. In general, the natural fish microflora tends to reflect the microbial communities of the surrounding waters. The gastrointestinal tract and gills typically yield high bacterial counts in fish, but these are affected by water quality and feed. Fish such as rainbow trout raised in clean and cold waters have lower bacterial counts than fish farmed or caught in warm waters (Mishra, 2022).

Conclusion

The proximate composition of rainbow trout varies according to the water quality of the environment where the fish live, the sex of the fish, the content of the fish feed, season, reproductive period, size group, and whether diploid or triploid. The crude protein content of rainbow trout ranged between 16.69 and 32.58%, moisture content ranged between 65.13 and 77.70%, lipid content ranged between 1.20 and 11.90%, and crude ash content ranged between 0.94 and 3.74%. The shelf life of rainbow trout depends on the fish feed content, muscle nutrient content of the fish, antioxidant substances used during processing in processing plants, and storage conditions.

References

Abedi E, Sahari MA. Long-chain polyunsaturated fatty acid sources and evaluation of their nutritional and functional properties. *Food Science and Nutrition* (2014) 2: 443e463.

Ackman RG. Seafood lipids and fatty acids. *Food Reviews International* (1990) 6(4): 617-646.

Akhtar P, Gray JL, Gornaa EA, Booren AM. Effect of dietary components and surface application of oleoresin rosemary on lipid stability of rainbow trout (*Oncorhynchus mykiss*) muscle during refrigerated and frozen storage. *Journal of Food Lipids* (1998) 5: 43-58.

Aksoy A, Sezer Ç. Combined use of laurel essential oil and vacuum packing to extend the shelf-life of rainbow trout (*Oncorhynchus mykiss*) fillets. *Kafkas Üniversitesi Veteriner Fakültesi Dergisi* (2019) *25*(6).

Asghari L, Zeynali F, Sahari MA. Effects of boiling, deep-frying, and microwave treatment on the proximate composition of rainbow trout fillets: Changes in fatty acids, total protein, and minerals. *Journal of Applied Ichthyology* (2013) *29*(4): 847-853.

Barkhori-Mehni S, Khanzadi S, Hashemi M, Azizzadeh M, Keykhosravy K. The Combined Effects of Bio-Components and Alginate Coating on Chemical and Sensory Quality of Chill-Stored Rainbow Trout Fillets. *Journal of Aquatic Food Product Technology* (2022) 1-13.

Bazargani-Gilani B. Activating sodium alginate-based edible coating using a dietary supplement for increasing the shelf life of rainbow trout fillet during refrigerated storage (4 ± 1C). *Journal of Food Safety* (2018) *38*(1): e12395.

Besharat M, Rajabi Islami H, Soltani M, Abdolmajid Mousavi S. Effect of different levels of nanoliposome-coated astaxanthin on growth performance, body proximate composition, liver enzyme activity and pigmentation of rainbow trout (*Oncorhynchus mykiss*). *Aquaculture Research* (2021) *52*(10): 5069-5077.

Bonilla AC, Sveinsdottir K, Martinsdottir E. Development of Quality Index Method (QIM) scheme for fresh cod (*Gadus morhua*) fillets and application in shelf life study. *Food control*, (2007) 18(4): 352-358.

Caballero MJ, Obach A, Rosenlund G, Montero D, Gisvold M, Izquierdo MS. Impact of different dietary lipid sources on growth, lipid digestibility, tissue fatty acid composition and histology of rainbow trout, *Oncorhynchus mykiss. Aquaculture* (2002) *214*(1-4): 253-271.

Celik M, Gökce MA, Başusta N, Kucukgulmez A, Taşbozan O, Tabakoğlu ŞS. Nutritional quality of rainbow trout (*Oncorhynchus mykiss*) caught from the Atatürk Dam Lake in Turkey. *Journal of Muscle Foods* (2008) *19*(1): 50-61.

Chaiyapechara S, Liu KK, Barrows FT, Hardy RW, Dong FM. Proximate composition, lipid oxidation, and sensory characteristics of fillets from rainbow trout *Oncorhynchus mykiss* fed diets containing 10% to 30% lipid. *Journal of the World Aquaculture Society* (2003) *34*(3): 266-277.

Dikici A, Özpolat E, BozatlI SB, Koluman A, Patir B, Çalıcıoğlu M. Investigating the effect of decontaminants on microbiological and chemical properties of rainbow trouts. *Turkish Journal of Veterinary & Animal Sciences* (2021) *45*(4): 691-699.

Dumas A, De Lange CF, France J, Bureau DP. Quantitative description of body composition and rates of nutrient deposition in rainbow trout (*Oncorhynchus mykiss*). *Aquaculture* (2007) *273*(1): 165-181.

Durmus M. The effects of nanoemulsions based on citrus essential oils (orange, mandarin, grapefruit, and lemon) on the shelf life of rainbow trout (*Oncorhynchus mykiss*) fillets at 4 ± 2C. *Journal of Food Safety* (2020) *40*(1): e12718.

Ehira S, Uchiyama H. Determination of fish freshness using the K value and comments on some other biochemical changes in relation to freshness. In: Kramer, D. E., Liston, J. (Eds.), *Seafood Quality Determination.* Elsevier Science, BV, Amsterdam (1986), 185–207.

Ehsani A, Jasour MS. Improvement of lipid stability of refrigerated rainbow trout (*Oncorhynchus mykiss*) fillets by pre-storage α-tocopherol acetate dipping treatment. In *Veterinary research forum* (2012) Vol. 3, No. 4, p. 269.

Erümit PÇ, Yıldız PO. Determination of Shelf Life of Rainbow Trout (*Oncorhynchus mykiss*) Fillets Cooked at Different Combinations of Temperatures Using the Sous Vide Technique. *Marine Science and Technology Bulletin* (2022) 11(2): 221-230.

Fallah AA, Siavash Saei-Dehkordi S, Nematollahi A. Comparative assessment of proximate composition, physicochemical parameters, fatty acid profile and mineral content in farmed and wild rainbow trout (*Oncorhynchus mykiss*). *International Journal of Food Science & Technology* (2011) *46*(4): 767-773.

FAO. *The State of World Fisheries and Aquaculture 2022. Towards Blue Transformation.* (2022). Rome, FAO.

Fıçıcılar BB, Gençcelep H. Influence of Nisin and Lysozyme on the Shelf Life of Hot-smoked Rainbow Trout Fillets (*Oncorhynchus mykiss*) during Storage at 4 C. *Philippine Journal of Science* (2021) *150*(3): 895-906.

Fonseca-Madrigal J, Karalazos V, Campbell PJ, Bell JG, Tocher DR. Influence of dietary palm oil on growth, tissue fatty acid compositions, and fatty acid metabolism in liver and intestine in rainbow trout (*Oncorhynchus mykiss*). *Aquaculture Nutrition* (2005) *11*(4): 241-250.

Frangos L, Pyrgotou N, Giatrakou V, Ntzimani A, Isavvaidis N. Combined effects of salting, oregano oil and vacuum-packaging on the shelf-life of refrigerated trout fillets. *Food Microbiology* (2010) 27:115–121.

Gokoglu N, Yerlikaya P, Cengiz E. Effects of cooking methods on the proximate composition and mineral contents of rainbow trout (*Oncorhynchus mykiss*). *Food chemistry* (2004) *84*(1): 19-22.

Gomes, E. F., Corraze, G., & Kaushik, S. Effects of dietary incorporation of a co-extruded plant protein (rapeseed and peas) on growth, nutrient utilization and muscle fatty acid composition of rainbow trout (*Oncorhynchus mykiss*). *Aquaculture* (1993) *113*(4): 339-353.

Güroy D, Güroy B, Merrifield DL, Ergün S, Tekinay AA, Yiğit M. Effect of dietary Ulva and Spirulina on weight loss and body composition of rainbow trout, *Oncorhynchus mykiss* (Walbaum), during a starvation period. *Journal of animal physiology and animal nutrition* (2011) *95*(3): 320-327.

Haliloğlu, H. İ., Bayır, A., Sirkecioğlu, A. N., Aras, N. M., & Atamanalp, M. Comparison of fatty acid composition in some tissues of rainbow trout (*Oncorhynchus mykiss*) living in seawater and freshwater. *Food Chemistry*, (2004) *86*(1): 55-59.

Hamzeh A, Rezaei, M. The effects of sodium alginate on quality of rainbow trout (*Oncorhynchus mykiss*) fillets stored at 4±2 C. *Journal of Aquatic Food Product Technology* (2012) *21*(1): 14-21.

Hashemi M, Pourmousavi FS, Mohajer F, Noori SMA. Impacts of Nano-Gelatin Coating Containing Thymol and Nisin on Chemical Quality Indices of Rainbow Trout Fillets Stored at 4° C. *Jundishapur Journal of Natural Pharmaceutical Products*, (2022) 17(3):e122177.

Hosseinzadeh Sahafi H, Masaeli S, Alizadeh M, Negarestan H, Naji T. *A study on growth parameters, blood factors and proximate composition of rainbow trout (Oncorhynchus mykiss) cultured in underground brackish and freshwater.* (2013) https://doi.org/10.1080/87559129009540896.

Hunt AO, Berkoz M, Ozkan F, Yalin S, Ercen Z, Erdogan E, Gunduz SG. Effects of organic selenium on growth, muscle composition, and antioxidant system in rainbow trout. *Israeli Journal of Aquaculture-Bamidgeh* (2011) *63*(562), 10.

Huss HH. Quality and quality changes in fresh fish - 4. Chemical composition. *Journal of AOAC International* (1995) 75(3): 488-506.

Islam MN, Joadder MAR. Seasonal variation of the proximate composition of freshwater Gobi, *Glossogobius giuris* (Hamilton) from the River Padma. *Pakistan Journal of Biological Sciences* (2005) 8(4): 532-536.

Johansson L, Kiessling A, Kiessling KH, Berglund L. Effects of altered ration levels on sensory characteristics, lipid content and fatty acid composition of rainbow trout (*Oncorhynchus mykiss*). *Food Quality and Preference* (2000) *11*(3): 247-254.

Köse I. Yildiz M. Effect of diets containing sesame oil on growth and fatty acid composition of rainbow trout (*Oncorhynchus mykiss*). *Journal of Applied Ichthyology* (2013) *29*(6): 1318-1324.

Mancini S, Medina I, Iaconisi V, Gai F, Basto A, Parisi G. Impact of black soldier fly larvae meal on the chemical and nutritional characteristics of rainbow trout fillets. *Animal* (2018) *12*(8): 1672-1681.

Mehdizadeh T, Tajik H, Jafarie S, Kaboudari A. Effect of Salvia officinalis L. extract on chemical, microbial, sensory and shelf life of rainbow trout fillet. *Food Science and Biotechnology* (2019) *28*(5): 1499-1506.

Mexis SF, Chouliara E, Kontominas MG. Combined effect of an oxygen absorber and oregano essential oil on shelf life extension of rainbow trout fillets stored at 4°C. *Food Microbiology* (2009) 26:598–605.

Mishra R. *Handbook on Fish Processing and Preservation*. CRC Press (2022).

Nazari K, Shamsaie M, Eila N, Kamali A, Sharifpour I. The effects of different dietary levels of organic and inorganic selenium on some growth performance and proximate composition of juvenile rainbow trout (*Oncorhynchus mykiss*). *Iranian Journal of Fisheries Sciences* (2017) *16*(1): 238-251.

Nowzari F, Shábanpour B, Ojagh SM. Comparison of chitosan–gelatin composite and bilayer coating and film effect on the quality of refrigerated rainbow trout. *Food chemistry* (2013) *141*(3): 1667-1672.

Nurhan, U. Change in proximate, amino acid and fatty acid contents in muscle tissue of rainbow trout (*Oncorhynchus mykiss*) after cooking. *International journal of food science & technology* (2007) *42*(9): 1087-1093.

Ojagh SM, Rezaei M, Razavi SH, Hosseini SMH. Effect of chitosan coatings enriched with cinnamon oil on the quality of refrigerated rainbow trout. *Food Chemistry* (2010) 120: 193–198.

Öz M, Dikel S, Durmuş M, Özoğul, Y. Effects of black cumin oil (Nigella sativa) on sensory, chemical and microbiological properties of rainbow trout during 23 days of storage at 2±1 C. *Journal of Aquatic Food Product Technology* (2017). *26*(6): 665-674.

Öz M, Dikel S, Durmuş M, Özşahinoğlu İ, Mumoğullarında P. Çörek otu (Nigella sativa, L) yağının Gökkuşağı Alabalığı (*Oncorhynchus mykiss*)'nın depolanması esnasında yağ asidi değişimine etkisi [Effect of Black Cumin (Nigella sativa, L) oil on fatty acid change during storage of Rainbow Trout (*Oncorhynchus mykiss*)]. *Nevşehir Bilim ve Teknoloji Dergisi* (2015) *4*(1), 57-68.

Öz M, Inanan B, Dikel, S. Yem Kaynaklı Borun Gökkuşağı Alabalığı (*Oncorhynchus mykiss*) Yağ Asidi Profiline Etkisi [Effect of Feed Source Boron on Fatty Acid Profile of Rainbow Trout (Oncorhynchus mykiss)]. *Avrupa Bilim ve Teknoloji Dergisi*, (2021) 31: 188-192.

Öz M, Ucak İ, Nayik GA. PUFA and MUFA. In *Nutraceuticals and Health Care* Academic Press (2022) pp. 199-215.

Öz M. Dikel S. Body compositions and fatty acid profile of *Salmo Trutta Macrostigma* caught from Korkun Brook. *Advan. Zool. Bot* (2015a) *3*(4): 190-192.

Öz M. Dikel S. Comparison of body compositions and fatty acid profiles of farmed and wild rainbow trout (*Oncorhynchus mykiss*). *Food Sci. Technol* (2015b) *3*: 56-60.

Öz M. Effects of garlic (*Allium sativum*) supplemented fish diet on sensory, chemical and microbiological properties of rainbow trout during storage at− 18 C. *LWT* (2018) *92*, 155-160.

Öz M. Tatil T. Dikel S. Effects of boric acid on the growth performance and nutritional content of rainbow trout (*Oncorhynchus mykiss*). *Chemosphere* (2021) 272, 129895.

Öz M. Dikel S. Effect of garlic (Allium sativum)-supplemented diet on growth performance, body composition and fatty acid profile of rainbow trout (*Oncorhynchus mykiss*. *Cellular and Molecular Biology*, (2022) 68(2).

Öz M. Nutrition and gender effect on body composition of rainbow trout (*Oncorhynchus mykiss*). *Journal of Advances in VetBio Science and Techniques* (2016) *1*(1): 20-25.

Ozogul Y, Yuvka İ, Ucar Y, Durmus M, Kösker AR, Öz M, Ozogul F. Evaluation of effects of nanoemulsion based on herb essential oils (rosemary, laurel, thyme and sage) on sensory, chemical and microbiological quality of rainbow trout (*Oncorhynchus mykiss*) fillets during ice storage. *Lwt* (2017) *75*: 677-684.

Özyurt G, Özkütük AS, Şimşek A, Yeşilsu AF, Ergüven M. Quality and shelf life of cold and frozen rainbow trout (*Oncorhynchus mykiss*) fillets: effects of fish protein-based biodegradable coatings. *International Journal of Food Properties* (2015) *18*(9): 1876-1887.

Peiretti PG, Gai F, Ortoffi M, Aigotti R, Medana C. Effects of rosemary oil (Rosmarinus officinalis) on the shelf-life of minced rainbow trout (*Oncorhynchus mykiss*) during refrigerated storage. *Foods* (2012) *1*(1): 28-39.

Pezeshk S, Rezaei M, Hosseini H. Effects of turmeric, shallot extracts, and their combination on quality characteristics of vacuum-packaged rainbow trout stored at 4 ± 1 C. *Journal of food science* (2011) *76*(6): M387-M391.

Pyrgotou N, Giatrakou V, Ntzimani A, Savvaidis IN. Quality assessment of salted, modified atmosphere packaged rainbow trout under treatment with oregano essential oil. *Journal of Food Science* (2010) 75: 406-411.

Rasmussen RS, Ostenfeld TH, Ronsholdt B, McLean E. Manipulation of end-product quality of rainbow trout with finishing diets. *Aquaculture Nutrition* (2000) *6*(1): 17-24.

Rezaei M, Hosseini SF. Quality assessment of farmed rainbow trout (*Oncorhynchus mykiss*) during chilled storage. *Journal of Food science* (2008) *73*(6): H93-H96.

Robertson R, Guihéneuf F, Schmid M, Stengel DB, Fitzgerald G, Ross P, Stanton C. *Algae-derived polyunsaturated fatty acids: implications for human health. Polyunsaturated fatty acids: Sources, antioxidant properties and health benefits*. USA: Nova science publishers (2013).

Rønsholdt B, McLean E. Effect of growth hormone and salbutamol on growth performance, fillet proximate composition and pigmentation of rainbow trout (*Oncorhynchus mykiss*). *Aquaculture* (2004) *229*(1-4): 225-238.

Rustan AC, Drevon CA. Fatty acids: Structures and properties. In *Encyclopedia of life sciences* (pp. 1e7). John Wiley & Sons, Ltd (2005).

Ryder JM, Fletcher GC, Stec MG, Seelye RJ. Sensory, microbiological and chemical changes in hoki stored in ice. *Int. J. Food Sci. Technol.* (1993) 28: 169–180.

Sabetian M, Delshad ST, Moini S, Islami HR, Motalebi A. Identification of fatty acid content, amino acid profile and proximate composition in rainbow trout (*Oncorhynchus mykiss*). *Journal of American Science*, (2012) *8*(4): 670-677.

Savary-Auzeloux I, Mosoni L, Rémond D, Dardevet D, Boirie Y, Stéphane Walrand S, Guillet C. *Protein digestion, absorption and metabolism.* (2014). https://www.researchgate.net/publication/260626399.

Serpek B, Kalaycıoglu L. Biyokimya. Ankara: *Nobel Yayin Dagitim* (2000).

Shadman S, Hosseini SE, Langroudi HE, Shabani S. Evaluation of the effect of a sunflower oil-based nanoemulsion with Zataria multiflora Boiss. essential oil on the physicochemical properties of rainbow trout (*Oncorhynchus mykiss*) fillets during cold storage. *LWT-Food Science and Technology* (2017) *79*: 511-517.

Taşbozan O, Gökçe MA, Erbaş C. The effect of different growing conditions to proximate composition and fatty acid profiles of rainbow trouts (*Oncorhynchus mykiss*). *Journal of applied animal research* (2016) *44*(1): 442-445.

Tokur B. The effect of different cooking methods on proximate composition and lipid quality of rainbow trout (*Oncorhynchus mykiss*). *International journal of food science & technology* (2007) *42*(7), 874-879.

Trung TS, Tran HV, Le MH, Ky PX, Brown PB, Van Ngo M. Growth performance, haematological parameters and proximate composition of rainbow trout *Oncorhynchus mykiss* fed varying dietary levels of protein hydrolysate from heads of Penaeus monodon shrimp processing industry. *Regional Studies in Marine Science* (2022) *55*: 102643.

Uçak İ, Öz M, Maqsood S. Products based on omega-3 polyunsaturated fatty acids and health effects. In *The Role of Alternative and Innovative Food Ingredients and Products in Consumer Wellness* (2019 pp. 197-212). Academic Press.

Wang X, Han Y. Comparison of the proximate composition, amino acid composition and growth-related muscle gene expression in diploid and triploid rainbow trout (*Oncorhynchus mykiss*) muscles. *J. Elem.* (2017) *22*(4): 1179-1191.

Yavuzer E, Özogul F, Özogul Y. Impact of icing with potato, sweet potato, sugar beet, and red beet peel extract on the sensory, chemical, and microbiological changes of rainbow trout (*Oncorhynchus mykiss*) fillets stored at (3 ± 1°C). *Aquaculture International* (2020) *28*(1): 187-197.

Yeşilayer N, Genç N. Comparison of proximate and fatty acid compositions of wild brown trout and farmed rainbow trout. *South African Journal of Animal Science* (2013) *43*(1): 89-97.

Yıldız M, Eroldoğan TO, Ofori-Mensah S, Engin K, Baltacı MA. The effects of fish oil replacement by vegetable oils on growth performance and fatty acid profile of rainbow trout: Re-feeding with fish oil finishing diet improved the fatty acid composition. *Aquaculture* (2018) *488*: 123-133.

Yıldız PO, Yangılar F. Effects of different whey protein concentrate coating on selected properties of rainbow trout (*Oncorhynchus mykiss*) during cold storage (4 C). *International Journal of Food Properties* (2016) *19*(9): 2007-2015.

Zare M, Mirzakhani MK, Stejskal V. Growth, body proximate composition and selected blood parameters of rainbow trout Onchorhynchus mykiss fingerlings fed on a diet supplemented with nettle Urtica dioica and tarragon *Artemisia dracunculus*. *Aquaculture Research* (2021) *52*(11): 5691-5702.

Zhaleh S, Shahbazi Y, Shavisi N. Shelf-Life Enhancement in Fresh and Frozen Rainbow Trout Fillets by the Employment of a Novel Active Coating Design. *Journal of food science* (2019) *84*(12): 3691-3699.

Contents of Earlier Volumes

Advances in Marine Biology. Volume 5

Advances in Marine Biology. Volume 4

Advances in Marine Biology. Volume 3

Advances in Marine Biology. Volume 2

Advances in Marine Biology. Volume 1

Index

P

R

S

T

W

Z